SPACE SCIENCE, EXPLORATION AND POLICIES

GLOBAL POSITIONING SYSTEMS

SPACE SCIENCE, EXPLORATION AND POLICIES

Additional books in this series can be found on Nova's website at:

https://www.novapublishers.com/catalog/index.php?cPath=23_29&seriesp=Space+
Science%2C+Exploration+and+Policies

Additional e-books in this series can be found on Nova's website at:

https://www.novapublishers.com/catalog/index.php?cPath=23_29&seriespe=Space+
Science%2C+Exploration+and+Policies

SPACE SCIENCE, EXPLORATION AND POLICIES

GLOBAL POSITIONING SYSTEMS

VIGGO ASPHAUG
AND
ELIAS SØRENSEN
EDITORS

Nova Science Publishers, Inc.
New York

Copyright © 2010 by Nova Science Publishers, Inc.

All rights reserved. No part of this book may be reproduced, stored in a retrieval system or transmitted in any form or by any means: electronic, electrostatic, magnetic, tape, mechanical photocopying, recording or otherwise without the written permission of the Publisher.

For permission to use material from this book please contact us:
Telephone 631-231-7269; Fax 631-231-8175
Web Site: http://www.novapublishers.com

NOTICE TO THE READER

The Publisher has taken reasonable care in the preparation of this book, but makes no expressed or implied warranty of any kind and assumes no responsibility for any errors or omissions. No liability is assumed for incidental or consequential damages in connection with or arising out of information contained in this book. The Publisher shall not be liable for any special, consequential, or exemplary damages resulting, in whole or in part, from the readers' use of, or reliance upon, this material. Any parts of this book based on government reports are so indicated and copyright is claimed for those parts to the extent applicable to compilations of such works.

Independent verification should be sought for any data, advice or recommendations contained in this book. In addition, no responsibility is assumed by the publisher for any injury and/or damage to persons or property arising from any methods, products, instructions, ideas or otherwise contained in this publication.

This publication is designed to provide accurate and authoritative information with regard to the subject matter covered herein. It is sold with the clear understanding that the Publisher is not engaged in rendering legal or any other professional services. If legal or any other expert assistance is required, the services of a competent person should be sought. FROM A DECLARATION OF PARTICIPANTS JOINTLY ADOPTED BY A COMMITTEE OF THE AMERICAN BAR ASSOCIATION AND A COMMITTEE OF PUBLISHERS.

LIBRARY OF CONGRESS CATALOGING-IN-PUBLICATION DATA

Asphaug, Viggo.
 Global positioning systems / Viggo Asphaug and Elias Sørensen.
   p. cm.
 Includes indexes.
 ISBN 978-1-60741-012-6 (hardcover)
 1. Global Positioning System. I. Sørensen, Elias. II. Title.
 G109.A76 2009
 910.285--dc22

                        2009040780

Published by Nova Science Publishers, Inc. ✦ New York

CONTENTS

Preface		**vii**
Chapter 1	Higher Order Ionospheric Errors in Modernized GPS and Future Galileo Systems *Mohammed M. Hoque and Norbert Jakowski*	**1**
Chapter 2	Highway Geometry Determination from GPS Data *María Castro, Luis Iglesias, Roberto Rodríguez-Solano and José A. Sánchez*	**29**
Chapter 3	How Location Performance Indexes of GPS Radio Collar Reflect Location Error in Mount Fuji, Central Japan *Zhaowen Jiang, Seiki Takatsuki, Masahiko Kitahara and Mikio Sugita*	**43**
Chapter 4	"Application of a Geographical Information System (GIS) and the Global Positioning System (GPS) to Dengue Virus Vector: *Aedes* Mosquitoes Distribution in an Epidemic Area of Thailand", A Technical comment *Viroj Wiwanitkit*	**57**
Chapter 5	Accurate GPS-Based Guidance of Agricultural Vehicles Operating on Slippery Ground *Benoit Thuilot, Roland Lenain, Philippe Martinet and Christophe Cariou*	**61**
Chapter 6	Global Positioning System Constraints on Plate Kinematics in the Southern Alps at the Nubia-Eurasia Boundary *M. Bechtold, D. Zuliani, P. Fabris, D.C. Tanner and M. Battaglia*	**117**
Chapter 7	Estimation of Regional Stress Increment Distribution Using GPS Array Data *Muneo Hori, Takeshi Iinuma and Teruyuki Kato*	**157**
Chapter 8	GPS-based Optimal FIR Filtering and Steering of Clock Errors *Yuriy Shmaliy*	**171**
Index		**239**

PREFACE

The Global Positioning System (GPS) is a global navigation satellite system (GNSS) developed by the United States Department of Defense and managed by the United States Air Force 50th Space Wing. It is the only fully functional GNSS in the world, can be used freely by anyone, anywhere, and is often used by civilians for navigation purposes. It uses a constellation of between 24 and 32 medium Earth orbit satellites that transmit precise radiowave signals, which allow GPS receivers to determine their current location, the time, and their velocity. Since it became fully operational on April 27, 1995, GPS has become a widely used aid to navigation worldwide, and a useful tool for map-making, land surveying, commerce, scientific uses, tracking and surveillance, and hobbies such as geocaching. Also, the precise time reference is used in many applications including the scientific study of earthquakes.

This new book gathers the latest research from around the globe in this dynamic field.

Modernized GPS and future Galileo systems will transmit a third frequency allowing to cancel out the first and second order ionospheric terms in the refractive index by double differencing carrier phases or code pseudoranges. However, differential bending of the signal and third order ionospheric term are not fully removed in this approach. Chapter 1 estimates the magnitude of various higher order ionospheric errors using a large number of ionospheric vertical profiles reconstructed from CHAMP-GPS radio occultation measurements. The authors investigation shows that triple-frequency residual range errors become significant at low elevation angles ($< 15°$) and at high level of total electron content. The triple-frequency residual range error reaches up to 1 and 4 cm at $5°$ elevation angle in carrier-phase and pseudo-range measurements, respectively. Different approximation formulas have been discussed to mitigate higher order ionospheric errors in operating satellite positioning systems.

The Galileo system will allow four frequencies to be used for higher order ionospheric corrections. It has been found that ionospheric effects are successfully removed in quadruple-frequency measurements, i.e., no ionospheric correction is required in the range estimation. However, quadruple-frequency measurements for the ionospheric correction are not practically useful since the measurement noise exceeds the ionospheric correction.

In many situations, highway geometry is unknown. Using GPS data to reconstruct highway geometry could be the best option in a variety of circumstances. Chapter 2 analyzes the performance of several GPS receivers in this task. Three different commercial receivers have been chosen for testing and performance comparison. The three devices reach submetric accuracy. The data have been collected in kinematic mode, with the GPS antennas attached to stable magnetic bases on the top of a vehicle, and post processed by differential corrections.

Tests have been carried out on a two-lane rural highway located on Spain. Using the GPS positions, a method for highway geometry reconstruction is presented. The obtained results offer versatile solutions for multiple civil engineering applications, including the incorporation of these data in GIS environments. In this way, analysis of traffic safety and highway consistency, and other kinds of studies, could be done easily.

In Chapter 3, the authors assessed the effects of habitat features with varying canopy closures, tree density, tree basal area, and topography on location performance of GPS radio collars (GPS3300, Lotek Engineering Inc., Ontario, Canada) around Mount Fuji, a single peak surrounded by wide and flat areas. The authors analyzed the differences of performance indexes (n = 1361) recorded by the GPS radio collars among different location status (unsuccessful fix, successful 2- dimensional (D), and 3-D location) and the relationships between indexes related location quality by successful location data. There were differences in all indexes (available satellite, location time, and location data dimension) among or between location status but not in position dilution of precision (PDOP). Available satellite number increased from unsuccessful to 2-D and to 3-D location. In contrast to available satellite, location time decreased. The differences between location indexes reflect location error and showed that 2-D location was greater than 3-D; however, the difference between 2- and 3-D was not reflected PDOP. Correlation analyses showed that available satellite number affected location time (r = -0.390), location data dimension (r = 0.639), PDOP (r = -0.362), and, finally, location error (r = -0.282). The location time, PDOP, and location error decreased with the increasing available satellite number. A higher satellite number yielded greater 3-D location. Meanwhile, small PDOP reflected less location error (r = 0.407) and shorter location time (r = 0.088). However, PDOP did not show a relation with the proportion of location data dimension (r = 0.042). Further, longer location time correlated with greater location error (r = 0.196) and less proportion of 3-D location (r = -0.442). The results confirmed the correlated relationships among performance indexes and the importance of available satellite number and PDOP reflected location quality of 2-D and 3-D independently, but not reflected location error synthetically.

In Thailand, dengue hemorrhagic fever is an important infectious disease. There are several outbreaks of this infection in each decade. This fact causes dengue hemorrhagic fever to be a major public health problem. As a rule, vector survey is a useful basic preventive technique for surveillance and control of mosquitoes. The author has ever published the previous report on dengue hemorrhagic fever using a global positioning technique. The objective of that quoted study is to apply the Geographical Information System (GIS) and the Global Positioning System (GPS) to the dengue virus vector, *Aedes* mosquitoes, distribution in an epidemic area of Thailand. In Chapter 4, the author provides a technical comment based on that report.

The development of automatic guidance systems for agricultural vehicles is receiving considerable attention from both researchers and manufacturers. The motivations in such automated devices are, on one hand, to reduce considerably the arduous driving task, and on the other hand, to improve the efficiency and the quality of the agronomic work carried out.

Such guidance devices require realtime vehicle localization on an unstructured area, such as agricultural fields. Nowadays, RTK GPS sensor appears as a very suitable sensor for these applications, since it can supply this information with a satisfactory centimeter accuracy at a high 10 *Hz* frequency, without requiring any preliminary equipment of the field.

In Chapter 5, it is demonstrated that very accurate curved path following can actually be achieved by agricultural vehicles, even on slippery ground, relying on a single RTK GPS sensor.

In a first step, in order to benefit from recent advances in Control Theory, sliding effects have been omitted, and therefore guidance laws have first been designed relying on a vehicle kinematic model. More precisely, taking advantage from structural properties of these models (they can be converted into a so-called chained form), curved path following has been achieved by designing a non-linear control law. Full-scale experiments reveal a very satisfactory guidance accuracy, except when the vehicle enters into sharp curves or when it moves on sloping fields.

In these two latter situations, guidance accuracy is damaged since the vehicle undergoes sliding effects. In agricultural applications, vehicle dynamic models appear untractable from the control design point of view. Therefore, it is here proposed to describe sliding effects as a structured perturbation acting on vehicle kinematic model. Adaptive Control framework, and more precisely Internal Model techniques, can then be used jointly with the above mentioned non-linear control law to reject sliding effects, when still preserving all the advantages of the previously designed non-linear guidance law.

Experiments demonstrate satisfactory guidance accuracy when the vehicle moves along a slope or when it executes sharp curves, excepted at their beginning or end. These transient guidance errors mainly ensue from delays introduced by the actuation device. Since the shape of the path to be followed is known, beginning/end of curves or slopes can be anticipated. Model Predictive Control framework is here used to provide such an anticipation. Satisfactory experimental results display the performances of the overall control scheme.

Chapter 6 presents and interprets deformation velocities from continuous Global Positioning System (GPS) observations at 42 sites around the Adriatic region (central Mediterranean) to investigate the active tectonics of the Nubia-Eurasia plate boundary in the northeast of Italy, the seismically most active region of the Southern Alps. In total, 1600 days of GPS observations, from June 2002 to November 2006, were processed using GAMIT/GLOBK. In detail, the authors present their processing scheme that is based on a three-step approach: i) estimation of station coordinates, atmospheric zenith delay, orbital and Earth orientation parameters using daily GPS observations and applying loose constraints to geodetic parameters; ii) combining, on a daily basis, the looselyconstrained solutions with loosely-constrained SOPAC solutions; and combining the daily combinations into monthly averages while adjusting the final $\sim_2$ to 1 by rescaling the daily h-files (method to account for white noise); and iii) defining the reference frame, determining the site velocities and estimating the error by including both the effect of white noise and random walk component (monument instability). The authors estimate the random walk component station by station using the "realistic sigma" algorithm of Herring (2003b).

All eight stations of the Friuli Regional Deformation Network (FReDNet), which form the focus of this chapter, present at least 2.5 years of data, and so exceed the minimum time span for a reliable determination of deformation velocity. The geodetic data indicate shortening of the crust in the region, with southern Friuli moving NNW towards northern Friuli at the relative speed of 1.6 to 2.2mm/yr. One of the eight processed FReDNet stations (i.e. CANV - Caneva), situated closely north to the active Polcenigo-Maniago Thrust, is moving faster than its expected long-term, geologic velocity. As there are no other plausible reasons for this discrepancy, the authors interpret this anomalous high velocity to be an indication of a locked thrust and predict that strain is accumulating to the north of it.

Chapter 7 presents an analysis method of GPS array data which is aimed at estimating stress increment filed that is associated with displacement or strain increment field that is observed by the GPS array. The method is based on an inverse analysis which finds self-equilibrating stress for a body with unknown stress-stress relation, when a distribution of strain is measured. With the assumptions of the plane stress state and no volumetric inelastic deformation for the incremental deformation of the Japanese Islands, the stress increment is computed from data which are measured by the GPS array that has been operating in Japan. It is shown that the stress increment is not uniform and that there are some regions which have sharp changes in the stress increment; these changes are associated with changes in observed strain increment even though the distribution of the estimated stress increment is different from that of the strain increment which is computed from the GPS array data. Some discussions are made for the usefulness and limitations of applying the present data analysis to the GPS array data.

Chapter 8 addresses novel results in optimal finite impulse response (FIR) filtering, prediction, and steering of the local clock time interval error (TIE) using the Global Positioning System (GPS) one pulse per second (1PPS) timing signals. Studies are motivated by permanently growing requirements for accuracy of timescales in different areas of applications of wire and wireless digital systems. Main limiters of accuracy here are the nonstationary random behavior of the clock TIE causing the model uncertainty, GPS time uncertainty due to different satellites in a view, and non Gaussian sawtooth noise induced by the commercially available GPS timing receivers owing to the principle of the 1PPS signal formation. Under such circumstances, the standard Kalman algorithm often becomes unstable and noisy, even if the sawtooth correction is used. The authors show that a better way is to use optimal FIR estimators, which are inherently bounded input/bounded output (BIBO) stable and more robust against temporary uncertainties and round-off errors. Among these estimators, simple unbiased FIR polynomial solutions have strong engineering features for slowly changing with time clock models. Moreover, they become optimal by large averaging horizons typically used in timekeeping. Theoretical studies of optimal FIR filtering are provided in detail. Applications are given for GPS-based measurements of crystal clocks.

In: Global Positioning Systems
Editors: V. Asphaug and E. Sørensen, pp. 1-28

ISBN: 978-1-60741-012-6
© 2010 Nova Science Publishers, Inc.

Chapter 1

HIGHER ORDER IONOSPHERIC ERRORS IN MODERNIZED GPS AND FUTURE GALILEO SYSTEMS

Mohammed M. Hoque[1] and Norbert Jakowski[2]
Institute of Communications and Navigation,
German Aerospace Center (DLR), Neustrelitz, Germany

Abstract

Modernized GPS and future Galileo systems will transmit a third frequency allowing to cancel out the first and second order ionospheric terms in the refractive index by double differencing carrier phases or code pseudoranges. However, differential bending of the signal and third order ionospheric term are not fully removed in this approach. This work estimates the magnitude of various higher order ionospheric errors using a large number of ionospheric vertical profiles reconstructed from CHAMP-GPS radio occultation measurements. Our investigation shows that triple-frequency residual range errors become significant at low elevation angles (< 15°) and at high level of total electron content. The triple-frequency residual range error reaches up to 1 and 4 cm at 5° elevation angle in carrier-phase and pseudo-range measurements, respectively. Different approximation formulas have been discussed to mitigate higher order ionospheric errors in operating satellite positioning systems. The Galileo system will allow four frequencies to be used for higher order ionospheric corrections. It has been found that ionospheric effects are successfully removed in quadruple-frequency measurements, i.e., no ionospheric correction is required in the range estimation. However, quadruple-frequency measurements for the ionospheric correction are not practically useful since the measurement noise exceeds the ionospheric correction.

---

1 E-mail adress: Mainul.Hoque@dlr.de. Phone: ++49 3981480 125. Kalkhorstweg 53, D-17235 Neustrelitz, Germany. Fax: ++49 3981480123.
2 E-mail address: Norbert.Jakowski@dlr.de. Phone: ++49 3981480151

Introduction

Ionosphere is a significant error source for users of space based navigation systems such as the GPS and future Galileo systems. For single-frequency GPS users, the signal delay due to refraction through the ionosphere corresponds to range errors of a few meters to tens of meters at the zenith [Klobuchar 1996] and it is the largest and most variable source of positioning error. Taking advantage of the dispersive nature of the ionosphere, dual-frequency users can eliminate the first order ionospheric term by differencing the signal at two frequencies. The remaining second and third order errors are found typically to be ~0-2 cm and ~0-2 mm at zenith, respectively, depending on the ionospheric and geomagnetic conditions [Bassiri and Hajj 1993]. A third frequency will be available from the modernized GPS and future Galileo systems. Therefore, triple-frequency users will be able to eliminate the first and second order ionospheric errors by double differencing code or phase measurements.

However, with a triple-frequency combination, the ionospheric error will not be removed completely. Higher order ionospheric errors caused by the differential bending of the signal and the third order ionospheric term are not fully removed in this approach. The bending of the ray path of Global Navigation Satellite Systems (GNSS) signals is caused by the ionospheric refractivity ($n \neq 1$) which mainly depends on the electron density distribution along the ray path. Bending effects lead to a deviation of the curved optical path from the free space path ($n = 1$). Due to the dispersive nature of the ionosphere, GPS signals at L1 (1575.42 MHz), L2 (1227.6 MHz) and L5 (1176.45 MHz) frequencies travel along different ray paths through the ionosphere. In each case the ray path is longer than the length of the free space path, in particular at low elevation angles. The corresponding excess path in addition to the free space path or true range may reach several centimeters at low elevations under high solar activity conditions [Jakowski et al. 1994]. Moreover, the total electron content will be different along the L1, L2 and L5 paths and causes an additional error in the range estimation.

This paper estimates the magnitude of the triple-frequency residual range errors due to total electron content difference, excess path length and third order ionospheric term. For this purpose, GPS signals have been traced through the ionosphere for a large number of electron density profiles reconstructed from CHAMP-GPS radio occultation measurements obtained at low and high solar activity conditions. CHAMP (CHAllenging Minisatellite Payload) is a small satellite mission especially for mapping magnetic and gravity fields of the Earth [Reigber et al., 2002]. It was launched on July 15, 2000 into a circular, near polar orbit with an initial altitude of 454 km. The onboard BlackJack GPS radio occultation receiver measures signals from setting GPS satellites using limb-pointed antenna. The GPS radio occultation measurements onboard CHAMP have been successfully used for retrieving vertical electron density profiles since 11 April 2001 [e.g., Jakowski 2005] which are available at http://w3swaci.dlr.de/.

In addition, different approaches have been discussed for higher order ionospheric corrections. The ionospheric effects in quadruple-frequency measurements have been investigated and briefly discussed the impact of the ionosphere-free combination on the measurement noise.

Higher Order Ionospheric Effects

The propagation of radio waves through the ionosphere/plasmasphere follows Fermat's principle, i.e., the optical distance L takes a minimum, where L is defined as the line integral between a satellite S and a receiver R along the ray path as

$$L = \int_S^R n\,ds \tag{1}$$

where n is the ionospheric phase refractive index and ds is the ray path element.

The excess path length d_I^{len} due to ray path bending of the signal is defined as an excess path in addition to the geometric path length (Euclidean line) between the transmitter and the receiver or true range ρ.

$$d_I^{len} = \int_S^R ds - \rho \tag{2}$$

Combining Eqs. (1) and (2), the range ρ can be expressed as

$$\rho = L + \int_S^R (1-n)\,ds - d_I^{len} \tag{3}$$

where $\int_S^R (1-n)ds$ is the phase delay experienced by the signal. Equation (3) separates the main types of ionospheric refraction effects. The phase delay is dominant over the excess path length d_I^{len} by about three orders of magnitude [e.g., Brunner and Gu 1991, Bassiri and Hajj 1993, Jakowski et al. 1994]; but the latter may be significant in high precision positioning especially at low elevation angles.

Ionospheric Refractive Index

The ionospheric refractive index is extensively studied by several authors [e.g., Lassen 1927, Hartree 1931, Appleton 1932]. Here, the refractive index expression given by the Appleton-Hartree formula is considered which is also known as Appleton-Lassen formula. At GNSS carrier frequencies the ionospheric plasma can be assumed to be collisionless. In this case, the Appleton-Hartree formula can be simplified as [Budden 1985, Davies 1990]

$$n^2 = 1 - \frac{2X(1-X)}{2(1-X) - Y^2 \sin^2 \Theta \pm \left[Y^4 \sin^4 \Theta + 4(1-X)^2 Y^2 \cos^2 \Theta \right]^{\frac{1}{2}}} \tag{4}$$

$$X = \frac{n_e e^2}{4\pi^2 \varepsilon_0 m} \cdot \frac{1}{f^2} = \frac{f_p^2}{f^2}$$

$$Y = \frac{eB}{2\pi m} \cdot \frac{1}{f} = \frac{f_g}{f}$$

(5)

The quantity n_e is the electron concentration, e and m are the electron charge and mass, respectively, ε_0 is the permittivity of the free space, f_p, f_g and f are the plasma, gyro and carrier frequencies, respectively, Θ is the angle between the Earth's magnetic field vector $\boldsymbol{B}$ and the propagation direction, and B is the magnitude of $\boldsymbol{B}$.

Since the ionospheric plasma is subjected to the geomagnetic field, it becomes anisotropic leading to a double refraction, which is indicated by the $\pm$ sign in Eq. (4). The wave with the upper (+) sign is usually called the 'ordinary' wave, whereas the lower (-) sign is related to the 'extraordinary' wave. The ordinary mode is left-hand circularly polarized, while the extra-ordinary mode is right-hand circularly polarized [Hartmann and Leitinger 1984].

The traditional way of deriving approximate expression of the refractive index is making assumptions about the angle Θ (e.g., approximations given by Tucker and Fannin 1968, Hartmann and Leitinger 1984). Later Brunner and Gu (1991), and Bassiri and Hajj (1993) used the order of magnitude of various terms of Eq. (4) in deriving a suitable expression.

Since the plasma frequency rarely exceeds 20 MHz [Klobuchar 1996] and gyro frequency is usually ≤ 1.4 MHz [Kelso 1975], the quantity X is of order of 10^{-5} and Y is of order of 10^{-3} at GNSS signal frequencies. Therefore, under certain assumptions (e.g., $Y << 2\cos\Theta(1 - X)/\sin^2\Theta$), the index of refraction can be expanded in inverse powers of frequency. The expansion of Eq. (4) up to the fourth inverse powers of frequency gives

$$n = 1 - \frac{1}{2}X \pm \frac{1}{2}XY\cos\Theta - \frac{1}{4}X\left[\frac{1}{2}X + Y^2\left(1 + \cos^2\Theta\right)\right]$$

(6)

Considering Eq. (5) this means:

$$n = 1 - \frac{f_p^2}{2f^2} \pm \frac{f_p^2 f_g \cos\Theta}{2f^3} - \frac{f_p^2}{4f^4}\left[\frac{f_p^2}{2} + f_g^2\left(1 + \cos^2\Theta\right)\right]$$

(7)

The second, third and fourth terms on the right hand side of Eq. (7) are proportional to the inverse square ($1/f^2$), inverse cube ($1/f^3$) and inverse quartic ($1/f^4$) powers of frequency, respectively. Equation (7) indicates that n is smaller than unity, which corresponds to a phase velocity greater than the speed of light in the vacuum (i.e., phase advance). Therefore, the integration of n along a signal path ($L = \int_S^R n ds$) between a receiver and a satellite leads to a range that is smaller than the geometric distance ρ by the amount of phase delay (advance).

Ionospheric Phase and Group Delays

As already mentioned, the ionospheric phase delay experienced by a signal passing through the ionosphere can be written as

$$d_I = \int_S^R (1-n)ds \tag{8}$$

When considering a group delay measurement, the phase refractive index n must be replaced by the group refractive index n_{gr}. The group refractive index n_{gr} can be expressed as a function of phase refractive index n by $n_{gr} = n + f(dn/df)$ or

$$n_{gr} = 1 + \frac{f_p^2}{2f^2} \mp \frac{f_p^2 f_g \cos\Theta}{f^3} + \frac{3f_p^2}{4f^4}\left[\frac{f_p^2}{2} + f_g^2\left(1+\cos^2\Theta\right)\right] \tag{9}$$

The upper (-) and lower (+) signs are related to the ordinary and extraordinary waves, respectively. It is seen that n_{gr} is greater than unity and related to a group velocity that is less than the speed of light. Therefore, the integration of n_{gr} along a signal path determines a range that is greater than ρ by the amount of group delay which can be written as

$$d_{Igr} = \int_S^R (n_{gr}-1)ds \tag{10}$$

Assuming a right hand circularly polarized wave as used by GPS and substituting n and n_{gr} values (Eqs. 7 and 9) in Eqs. (8) and (10), the ionospheric phase and group delay expressions can be written in terms of inverse powers of radio wave frequency as

$$d_I = \frac{p}{f^2} + \frac{q}{2f^3} + \frac{u}{3f^4} \tag{11}$$

$$d_{Igr} = \frac{p}{f^2} + \frac{q}{f^3} + \frac{u}{f^4} \tag{12}$$

$$p = \frac{K}{2}\int_{path} n_e ds = \frac{K \cdot TEC}{2} = \frac{K}{2}(TEC_{LoS} + \Delta TEC) \tag{13}$$

$$q = 2.2566 \times 10^{12} \int n_e B \cos\Theta \cdot ds \tag{14}$$

$$u = 2437 \int n_e^2 ds + 4.74 \times 10^{22} \int n_e B^2 (1+\cos^2\Theta)ds \tag{15}$$

where $K = e^2/(4\pi^2\varepsilon_0 m) = 80.6$ m³s⁻² and the terms p/f^2, $q/(2f^3)$ and $u/(3f^4)$ in Eq. (11) are the first, second and third order ionospheric phase delays, respectively. The corresponding group delays are p/f^2, q/f^3 and u/f^4, respectively. Comparing Eqs. (11) and (12), we see that the magnitude of the first order group delay is equal to the first order phase delay; however, the second and third order group delays are two and three times of the respective phase delays.

The integral $\int_{path} n_e ds$ along a signal path (i.e., curved path) is defined as the total electron content TEC and often measured in TEC units (1 TECU = 10^{16} electrons/m²). Due to the dispersive nature of the ionosphere, the ray path bending effects at two different frequencies are not equal; therefore, two GNSS signals travel along different ray paths through the ionosphere and the TEC along f_1 path will be different from that along f_2 path and also from that along the straight line of sight (LoS) path. Considering this, TEC in Eq. (13) is separated into TEC_{LoS} which is along the LoS and ΔTEC which is the difference between TECs along the curved path and the LoS. The second and third order effects are not considered in TEC estimation in Eq. (13). Therefore, the term ΔTEC represents TEC contribution due to ray path bending only.

Ionospheric Effects on GNSS Observables

The GNSS observables are ranges which are deduced from measured time or phase differences based on a comparison between received signals and receiver generated signals. Thus, the ranges are biased by satellite and receiver clock errors, instrumental biases and atmospheric effects. The GNSS carrier-phase (Φ) and pseudo-range (Ψ) at a selected frequency can be described by the observation equations in units of lengths as

$$\Phi = \rho + c(dt_{clk} - dT_{clk}) - d_I^* + d_A + \left(d_{MP}\right)_\Phi + dq + dQ + N\lambda + \varepsilon_\Phi \qquad (16)$$

$$\Psi = \rho + c(dt_{clk} - dT_{clk}) + d_{Igr}^* + d_A + \left(d_{MP}\right)_\Psi + dq + dQ + \varepsilon_\Psi \qquad (17)$$

where ρ is the geometric distance between the satellite and receiver, c is the velocity of light, dt_{clk} and dT_{clk} are the satellite and receiver clock errors, respectively, d_I^* and d_{Igr}^* are the ionospheric errors in carrier-phase and pseudo-range measurements, respectively, d_A is the atmospheric (tropospheric) delay, $(d_{MP})_\Phi$ and $(d_{MP})_\Psi$ are the errors due to multipath in carrier-phase and pseudo-range measurements, respectively , dq and dQ are the instrumental biases of the satellite and receiver, respectively, λ is the carrier wavelength, N is the integer carrier-phase ambiguity, and ε_Ψ and ε_Φ are random errors.

Different non dispersive terms which are independent of the frequency such as the geometric distance ρ, atmospheric delay and clock errors can be removed by differencing the signal at two frequencies f_1 and f_2. Since we confine our interest only on computing higher order ionospheric effects, the ambiguity term, multipath effect and the satellite and receiver instrumental biases have been neglected. The carrier-phase and code pseudo-range equations (Eqs. 16 and 17) can be simplified accordingly and written in terms of range and ionospheric errors as [Hoque and Jakowski 2007, 2008; Bassiri and Hajj 1993]

$$\Phi_i = \rho - d_i^* = \rho - d_i + d_i^{len} = \quad \rho - \frac{p_i}{f_i^2} - \frac{q}{2f_i^3} - \frac{u}{3f_i^4} + d_i^{len} \tag{18}$$

$$\Psi_i = \rho + d_{Igr}^* = \rho + d_{Igr} + d_i^{len} = \quad \rho + \frac{p_i}{f_i^2} + \frac{q}{f_i^3} + \frac{u}{f_i^4} + d_i^{len} \tag{19}$$

where the index i = 1, 2, 3... refers to the GNSS carrier frequencies. The GPS carrier frequencies of L1, L2 and L5 signals are f_1 = 1575.42, f_2 = 1227.6 and f_3 = 1176.45 MHz, respectively. As we see in Eqs. (18) and (19), ionospheric errors include both the phase or group delay and the excess path length term d_I^{len}.

The most part of the first order ionospheric delay, i.e., TEC_{LoS} term in Eq. (13) can be eliminated by constructing a linear combination of dual-frequency observables. However, the second and third order ionospheric terms and the errors due to the excess path length and TEC difference will not cancel out in this approach. Linear combinations of GPS L1-L2 and L1-L5 carrier phases provide:

$$\begin{aligned}
\Phi_1 f_1^2 - \Phi_2 f_2^2 &= \rho\left(f_1^2 - f_2^2\right) + \frac{K}{2}\left(\Delta TEC_2 - \Delta TEC_1\right) \\
&+ \frac{q}{2}\frac{\left(f_1 - f_2\right)}{f_1 f_2} + \frac{u}{3}\frac{\left(f_1^2 - f_2^2\right)}{f_1^2 f_2^2} - \left(f_2^2 d_2^{len} - f_1^2 d_1^{len}\right)
\end{aligned} \tag{20}$$

$$\begin{aligned}
\Phi_1 f_1^2 - \Phi_3 f_3^2 &= \rho\left(f_1^2 - f_3^2\right) + \frac{K}{2}\left(\Delta TEC_3 - \Delta TEC_1\right) \\
&+ \frac{q}{2}\frac{\left(f_1 - f_3\right)}{f_1 f_3} + \frac{u}{3}\frac{\left(f_1^2 - f_3^2\right)}{f_1^2 f_3^2} - \left(f_3^2 d_3^{len} - f_1^2 d_1^{len}\right)
\end{aligned} \tag{21}$$

where ΔTEC_1, ΔTEC_2 and ΔTEC_3 are the excess TEC in addition to the LoS TEC due to bending along f_1, f_2 and f_3 signal paths, respectively. We see that the second order term cancels out in the triple-frequency combination (using Eqs. 20 and 21).

$$\frac{1}{C}\left[A\left(\Phi_1 f_1^2 - \Phi_2 f_2^2\right) - B\left(\Phi_1 f_1^2 - \Phi_3 f_3^2\right)\right] = \rho \underbrace{-\left(\Delta s_{TEC}\right)_{tr} - \left(\Delta s_3\right)_{tr} + \left(\Delta s_{len}\right)_{tr}}_{(RRE)_{tr}} \tag{22}$$

In which

$$\left(\Delta s_{TEC}\right)_{tr} = \frac{K}{2C}\left[B\left(\Delta TEC_3 - \Delta TEC_1\right) - A\left(\Delta TEC_2 - \Delta TEC_1\right)\right] \tag{23}$$

$$\left(\Delta s_3\right)_{tr} = \frac{u}{3C}\frac{\left(f_2 - f_3\right)}{f_2 f_3} \tag{24}$$

$$\left(\Delta s_{len}\right)_{tr} = \frac{1}{C}\left[B\left(f_3^2 d_3^{len} - f_1^2 d_1^{len}\right) - A\left(f_2^2 d_2^{len} - f_1^2 d_1^{len}\right)\right] \tag{25}$$

$$\left.\begin{array}{l} A = \dfrac{f_1 f_2}{f_1 - f_2} \\[3mm] B = \dfrac{f_1 f_3}{f_1 - f_3} \\[3mm] C = f_1\left(f_2 - f_3\right)\left(f_1 + f_2 + f_3\right) \end{array}\right\} \tag{26}$$

where $(\Delta s_{TEC})_{tr}$ is the tripe-frequency residual error due to TEC difference along ray paths, $(\Delta s_3)_{tr}$ is the third order residual error and $(\Delta s_{len})_{tr}$ is the residual error due to excess path length. The residual range error $(RRE)_{tr}$, which is the sum of all residual errors, is the difference between the determined range and the true range (see Eq. 22).

Similarly, the pseudorange of code or group delay measurement is given by

$$\frac{1}{C}\left[A\left(\Psi_1 f_1^2 - \Psi_2 f_2^2\right) - B\left(\Psi_1 f_1^2 - \Psi_3 f_3^2\right)\right] = \rho \underbrace{+\left(\Delta s_{TEC}\right)_{tr} + 3\left(\Delta s_3\right)_{tr} + \left(\Delta s_{len}\right)_{tr}}_{\left(RREgr\right)_{tr}} \tag{27}$$

where the total residual range error is denoted by $(RREgr)_{tr}$.

Higher Order Effects Computation

To trace radio waves through the ionosphere a two-dimensional ray tracing program [Hoque and Jakowski 2008] has been developed. This computer program traces the path of radio waves through a user-specified model that describes the spatial electron density distribution n_e and the geomagnetic field B. The vertical electron density distribution is deduced from numerous vertical electron density profiles obtained from GPS radio occultation measurements on board CHAMP [e.g., Jakowski 2005]. Horizontal gradients of the ionospheric ionization are neglected by assuming a spherically layered ionosphere. The effect of the Earth's magnetic field on the radio wave propagation is taken into account by calculating geomagnetic field vectors at numerous points along incoming ray paths applying the IGRF model [Mandea and Macmillan 2000].

Knowing the vertical distribution of the refractive index deduced from the retrieved electron density profiles, the rays have been traced through the ionosphere using Bouguer's law (Snell's law in spherically layered propagation medium) of refraction. The refractive index equation in an anisotropic plasma has been solved by introducing a quartic equation known as Booker quartic [Booker 1949, page 272-274] as described by Budden (1985, page 74-79). To solve the homing-in problem Nelder-Mead [Nelder and Mead 1965] simplex algorithm is implemented in the ray tracing program. The differences between excess path lengths estimated by our ray tracing technique and Brunner and Gu (1991) results were found to be less than one millimetre even at 7.5° elevation and high vertical TEC of 455 TECU at GPS frequencies. Brunner and Gu used a three-dimensional ray tracing technique based on Hamiltonian differential equations [e.g., Haselgrove 1963, Jones and Stephenson 1975]. For details about the ray tracing technique the reader is referred to Hoque and Jakowski (2008).

Due to the low CHAMP orbit height the topside ionosphere/plasmasphere contribution from altitude above CHAMP has been modelled by an adaptive Chapman layer superposed by an exponential decay for the plasmasphere [Jakowski et al. 2002, Jakowski 2005]. More than four hundred thousands CHAMP profiles with global coverage are available till now. For ionospheric error estimations, from these we have selected about 29,658 and 20,058 profiles retrieved during high and low solar activity years 2002 and 2006, respectively. Extreme electron density profiles having vertical TECs less than 10 TECU and greater than 300 TECU are excluded. Selected profiles are plotted in Figure 1.

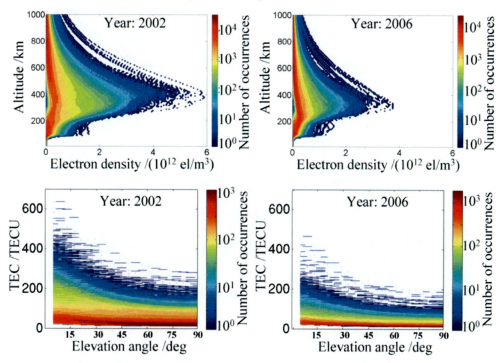

Figure 1. CHAMP based electron density profiles reconstructed during 2002 and 2006 (top panel) and corresponding TEC values at different elevation angles (bottom panel).

The color representation of the number of occurrences (see Figure 1) gives an opportunity to view a large number of electron density profiles in a single image. The vertical profiles up to 1000 km altitude are shown in the upper panel. The peak electron densities and their occurrence numbers are also visible in these images. We see that the maximum ionization is below 10^{12} electrons/m^3 for most of the profiles but in few cases it is close to 6×10^{12} electrons/m^3 during 2002.

The slant TEC estimates at different elevations are given in the lower panel of Figure 1. The total number of occurrences at a given elevation angle is equal to the total number of profiles considered. We see that TEC exceeds the 500 TECU level at low elevation angles (< 15°) during high solar activity conditions in 2002. The maximum vertical TECs at 90°

elevation angle are about 281 TECU (values > 300 TECU have been ignored in this study) and 186 TECU during 2002 and 2006, respectively. However, for most of the profiles TECs are well below these numbers. Considering this, in Figure 2, the root mean squared (RMS) value of TEC is plotted together with the maximum (MAX) value of TEC at different elevation angles. RMS estimates have been used in this study frequently. As an example, the RMS value of TEC can be written as

$$TEC_{RMS} = \sqrt{\frac{1}{m}\sum_{i=1}^{m} TEC_i^2} \qquad (28)$$

where m is the total number of elements.

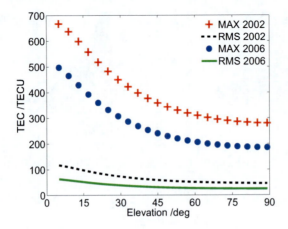

Figure 2. RMS and maximum estimates of TEC (for samples drawn in Figure 1).

It has been found that at zenith propagation or 90° elevation, the RMS TEC estimates are about 46 and 24 TECU during 2002 and 2006, respectively. At 5° elevation, they are about 118 and 63 TECU, respectively.

The electron density profile obtained from each CHAMP-GPS radio occultation corresponds to a defined geographic position on the Earth's surface. Ray tracing simulations take that position as the receiver position. Then satellite positions at GPS orbit height are derived knowing the receiver position and assuming a variety of receiver-to-satellite elevation and azimuth angles. The elevation angle is varied from 5-90° with a step size of 4° and in each case the azimuth angle is randomly selected from its range 0-360°. The azimuth angle is measured clockwise from the north, i.e., the geographic North Pole has an azimuth of 0° from every other point on the globe. Thus, at a certain elevation angle simulations are done for about thirty thousand (29,658) profiles for the year 2002 and about twenty thousand (20,058) profiles for the year 2006 (see Figure 1). The ray paths are traced for L1, L2 and L5 signals using the numerical ray tracing program. In each case, the higher order ionospheric terms such as the third order $(\Delta s_3)_{tr}$ term and ray path bending effects $(\Delta s_{TEC})_{tr}$ and $(\Delta s_{len})_{tr}$ are computed. Their maximum and RMS estimates are determined at different elevation angles. The results are plotted and discussed in the following sections.

Third Order Residual Error

The MAX and RMS estimates of the third order residual error $(\Delta s_3)_{tr}$ are plotted in Figure 3 as a function of the elevation angle.

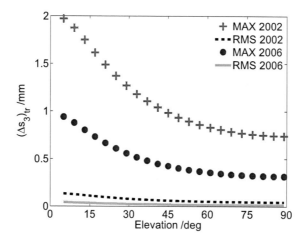

Figure 3. RMS and maximum estimates of $(\Delta s_3)_{tr}$ for years 2002 and 2006.

It has been found that at 5° elevation angles, $(\Delta s_3)_{tr}$ may reach up to the 2 mm level for the GPS L1-L2-L5 phase combination during times of high TEC. However, for most of the profiles, $(\Delta s_3)_{tr}$ is below 0.1 mm even at low elevation angles. At zenith $(\Delta s_3)_{tr}$ may reach up to about 0.7 mm and 0.3 mm during high and low solar activity years, respectively.

At zenith the RMS estimates are about 0.05 and 0.01 mm during high and low solar activity years 2002 and 2006, respectively. At 5° elevation angle they are about 0.13 and 0.04 mm, respectively.

Error due to TEC Difference

Due to the dispersive nature of the ionosphere L1, L2 and L5 signals travel along different ray paths through the ionosphere. Therefore, the integrated electron densities (TEC) along the ray paths are different. This causes an error in the triple-frequency range estimation which is defined here as $(\Delta s_{TEC})_{tr}$ (see Eq. 23). The maximum and RMS estimates of $(\Delta s_{TEC})_{tr}$ at different elevation angles are plotted in Figure 4.

We see that the maximum and RMS estimates of $(\Delta s_{TEC})_{tr}$ can be as big as 22 and 2 mm, respectively, at low elevation angles during high solar activity year 2002. During low solar activity year 2006 they are about 12 and 0.6 mm, respectively. Figure 4 shows that the maximum and RMS estimates of $(\Delta s_{TEC})_{tr}$ decrease rapidly with the increase of the elevation angle and disappear at zenith as expected.

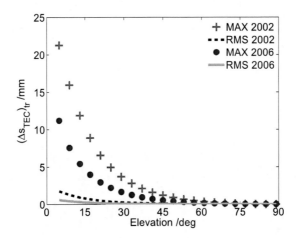

Figure 4. RMS and maximum estimates of $(\Delta s_{TEC})_{tr}$ for years 2002 and 2006.

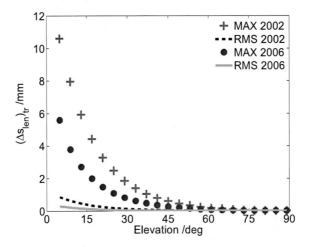

Figure 5. RMS and maximum estimates of $(\Delta s_{len})_{tr}$ for years 2002 and 2006.

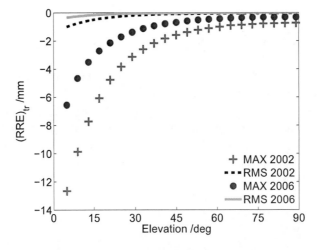

Figure 6. RMS and maximum estimates of $(RRE)_{tr}$ for years 2002 and 2006.

Error due to Excess Path Length

The MAX and RMS estimates of the excess path length error in the triple-frequency combination ($\Delta s_{len})_{tr}$ (see Eq. 25) are plotted in Figure 5. We see that the maximum and RMS estimates decrease rapidly with the increase of the elevation angle. At zenith propagation there is no ray path bending effect and there is no excess path length error. Simulation using CHAMP based profiles shows that $(\Delta s_{len})_{tr}$ can be as big as 11 and 6 mm during 2002 and 2006, respectively, at low elevation angles. However, the RMS estimates are below 1 mm.

Residual Range Error in the Phase Combination

The total ionospheric range errors in the triple-frequency phase combination $(RRE)_{tr}$, (see Eq. 22) are computed by adding the residual errors $(\Delta s_{TEC})_{tr}$, $(\Delta s_3)_{tr}$ and $(\Delta s_{len})_{tr}$. It has been found that the absolute $(RRE)_{tr}$ can reach up to 13 mm at low elevation angles during high solar activity year 2002, whereas it can be as high as 7 mm during low solar activity year 2006. At zenith these values can be as big as about 0.7 mm and 0.3 mm, respectively. However, in most cases the $(RRE)_{tr}$ is well below these numbers. The RMS estimates plotted in Figure 6 confirm this. The RMS estimates are found to be about 1 and 0.05 mm at 5° and 90° elevations, respectively, during 2002.

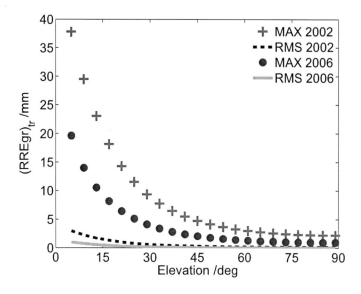

Figure 7. RMS and maximum estimates of $(RREgr)_{tr}$ for years 2002 and 2006.

Residual Range Error in the Code Combination

The residual range errors in the triple-frequency code combination $(RREgr)_{tr}$ (see Eq. 27) are computed, and their maximum and RMS estimates are plotted in Figure 7. It has been found that the $(RREgr)_{tr}$ can be as big as 38 mm and 20 mm during 2002 and 2006, respectively at 5° elevation. Comparing Figures 6 and 7, we see that the $(RREgr)_{tr}$ values are higher than the

(RRE)$_{tr}$ values. One reason is that the third order term $(\Delta s_3)_{tr}$ is three times higher in the code combination than in the phase combination. However, $(\Delta s_3)_{tr}$ is very small compared to $(\Delta s_{TEC})_{tr}$ and $(\Delta s_{len})_{tr}$ at low elevation angles. The other reason is that the excess path length term $(\Delta s_{len})_{tr}$ is additive to the other terms in the code combination, whereas it is subtractive in the phase combination (see Eqs. 22-27).

The RMS estimates are found to be about 4 and 0.14 mm at 5° and 90° elevations, respectively, during 2002.

Figures 6 and 7 show that during high solar activity year 2002, higher order ionospheric effects are significant for triple-frequency measurements and cannot be ignored if millimeter level accuracy is needed in the range estimation. In the following section, different approximation formulas are discussed for higher order corrections in the triple-frequency measurements.

Higher Order Effects Correction

One method to correct higher order ionospheric terms in the operating satellite positioning systems is the introduction of the empirical formulas which estimate the higher order terms using ionospheric parameters such as the TEC, maximum ionization N_m, its height h_m, and geometrical parameters such as the elevation β and azimuth angle.

Residual Error $(\Delta s_{TEC})_{tr}$ Correction

In our previous work [Hoque and Jakowski 2008], we investigated the range error due to TEC difference in the GPS dual-frequency measurements. To calculate the TEC difference along two GNSS signals, we derived an approximation based on analytical integration of ray path characteristics. For this, we used a single Chapman layer [Rishbeth and Garriott 1969] assumption of the ionosphere and its quasi-parabolic (QP) layer [Chen et al. 1990] representation.

The GNSS signal ray paths are largely separated (e.g., about 60 m at 7.5° elevation and vertical TEC of 138 TECU [Brunner and Gu 1991]) at and near the maximum ionization height, whereas their separation at regions of low ionization is small. As a result, the difference in TEC measurements along two signal paths mostly comes from the ionospheric region at and near the maximum ionization height. A QP layer can represent the profile n_e at and near the maximum ionization height well and gives an analytically integrable function for the TEC difference expression [Hoque and Jakowski 2008].

Although the approximation is derived for the dual-frequency correction, it can be used for the triple-frequency correction as well. The expression for $(\Delta s_{TEC})_{tr}$ is given in Eq. (23). The unknown components $\left(\Delta TEC_3 - \Delta TEC_1\right)$ and $\left(\Delta TEC_2 - \Delta TEC_1\right)$ can be calculated by the following formula.

$$\Delta TEC_j - \Delta TEC_i = TEC_j - TEC_i = \left(\frac{1}{f_j^2} - \frac{1}{f_i^2}\right)\left(I_1 + I_2 + I_3 + I_4 + I_5\right) \quad (29)$$

in which

$$I_1 = \frac{KU^2}{2a^3}\left[-\frac{a}{\sqrt{r^2-a^2}}-\frac{3}{2}\cos^{-1}\left(\frac{a}{r}\right)-\frac{a\sqrt{r^2-a^2}}{2r^2}\right]_{r_b}^{r_t}$$

$$I_2 = -\frac{KUV}{a^2}\left(\frac{2r^2-a^2}{r\sqrt{r^2-a^2}}\right)_{r_b}^{r_t}$$

$$I_3 = -\frac{K\left(2UW+V^2\right)}{2a}\left[\frac{a}{\sqrt{r^2-a^2}}+\cos^{-1}\left(\frac{a}{r}\right)\right]_{r_b}^{r_t} \qquad (30)$$

$$I_4 = -KVW\left(\frac{r}{\sqrt{r^2-a^2}}\right)_{r_b}^{r_t}$$

$$I_5 = -\frac{KW^2a^2}{2}\left(\frac{1}{\sqrt{r^2-a^2}}\right)_{r_b}^{r_t}$$

where $K = 80.6$ m^{3}s^{-2} and $a = r_0 \cos\beta_0$, r_0 is the radial distance from the Earth's centre to the receiver and β_0 is the elevation angle at the receiver. The quantities $U = -\left(N_m \cdot r_m^2 \cdot r_b^2\right)/y_m^2$, $V = 2N_m \cdot r_m \cdot r_b^2 / y_m^2$ and $W = N_m - N_m \cdot r_b^2 / y_m^2$ are coefficients of a QP layer. The QP layer is only valid for $r_b < r < r_t$ where $r_b = r_m - y_m$, $r_t = r_m r_b /\left(r_b - y_m\right)$, r is the radial distance from the Earth's centre, r_m is the radial distance from the Earth's centre to the height of maximum ionization, N_m is the value of maximum ionization and y_m is the layer semithickness. The TEC difference between two GNSS signals $TEC_j - TEC_i$ will be measured in electrons/m^2 if r_m and y_m are measured in meters, N_m is in electrons/m^3 and f is in Hertz. Then, the quantity $(\Delta s_{TEC})_{tr}$ will be measured in meters by Eq. (23). The detailed derivation of the above formula is given in Hoque and Jakowski (2008).

The analytical solution (Eqs. 29, 30) for determining the TEC difference is based on a single Chapman layer and its QP layer representation. However, the actual ionosphere cannot always be described by a single Chapman layer. Considering this, based on multi-layered Chapman profiles, an alternative correction formula is proposed for the excess TEC computation in Hoque and Jakowski (2008).

$$\Delta TEC_i = \frac{1.108 \times 10^{-3} \cdot TEC^2 \cdot \exp\left(-2.1844\beta\right)}{f_i^2 H \cdot \left(h_m\right)^{0.3}} \qquad (31)$$

where ΔTEC_i will be measured in TECU when ionospheric F_2 layer scale height H and peak ionization height h_m are in km, frequency f_i is in GHz, slant TEC is in TECU and elevation β is in radians. Now, the residual error $(\Delta s_{TEC})_{tr}$ can be estimated by Eq. (23) in conjunction with Eq. (31) in meters if ΔTEC_i is in electrons/m^2 and f in Hertz.

Equation (23) shows that $(\Delta s_{TEC})_{tr}$ depends on the signal frequency f_1, f_2, f_3, and TEC difference (e.g., $\Delta TEC_3 - \Delta TEC_1$) between signal paths. The TEC difference again depends on the signal frequency and ionospheric parameters (see Eqs. 29 or 31). Substituting ΔTEC_i

expression (Eq. 31) in Eq. (23), $(\Delta s_{TEC})_{tr}$ can be expressed as a function of dispersive (i.e., frequency dependent) and non-dispersive factors.

$$
\begin{aligned}
\left(\Delta s_{TEC}\right)_{tr} &= \frac{K}{2C} \cdot J \cdot \left[\frac{f_1 f_3}{(f_1 - f_3)} \left(\frac{1}{f_3^2} - \frac{1}{f_1^2} \right) - \frac{f_1 f_2}{(f_1 - f_2)} \left(\frac{1}{f_2^2} - \frac{1}{f_1^2} \right) \right] \\
&= \frac{K}{2 f_1 (f_2 - f_3)(f_1 + f_2 + f_3)} \cdot J \cdot \frac{(f_2 - f_3)}{f_2 f_3} \\
&= \frac{K}{2 f_1 f_2 f_3 (f_1 + f_2 + f_3)} \cdot J
\end{aligned}
\tag{32}
$$

in which

$$
J = \frac{1.108 \times 10^{-3} \cdot TEC^2 \cdot \exp(-2.1844\beta)}{H \cdot (h_m)^{0.3}}
\tag{33}
$$

where J is a frequency independent term and always positive. Equation (32) shows that $(\Delta s_{TEC})_{tr}$ will be positive independently of the relative magnitudes of f_1, f_2 and f_3 frequencies i.e., independent of frequency selections $f_1 > f_2 > f_3$ or $f_1 < f_2 < f_3$. However, its sign is negative in the phase combination (Eq. 22) and positive in the pseudo-range combination (Eq. 27).

Residual Error $(\Delta s_3)_{tr}$ Correction

To correct the third order ionospheric term, a correction formula based on analytical integration of a Chapman profile has been given in Hoque and Jakowski (2008). The unknown component u in the $(\Delta s_3)_{tr}$ expression (Eq. 24) can be expressed as

$$
u = 1602.75 \times N_m TEC
\tag{34}
$$

using formulas given in Hoque and Jakowski, where N_m is the maximum ionization and TEC is the total electron content. The quantity $(\Delta s_3)_{tr}$ will be measured in meters if N_m is measured in electrons/m^3, TEC is in electrons/m^2 and f in Hertz.

Since u (Eq. 34) is a non-dispersive term, substituting u in Eq. (24), $(\Delta s_3)_{tr}$ can be expressed in terms of dispersive and non-dispersive terms.

$$
\left(\Delta s_3\right)_{tr} = \frac{u}{3 f_1 f_2 f_3 (f_1 + f_2 + f_3)}
\tag{35}
$$

We see that $(\Delta s_3)_{tr}$ is positive independently of frequency selections. Its sign is negative in the phase combination (Eq. 22) and positive in the pseudo-range combination (Eq. 27).

Excess Path Length $(\Delta s_{len})_{tr}$ Correction

The expression for $(\Delta s_{len})_{tr}$ is given in Eq. (25). Unknown components d_1^{len}, d_2^{len} and d_3^{len}, which are the excess path lengths for the L1, L2 and L5 signals, respectively, can be calculated by the following formula [Hoque and Jakowski 2008].

$$d_i^{len} = \frac{7.5 \times 10^{-5} \cdot \exp(-2.13\beta) \cdot TEC^2}{f_i^4 H \cdot (h_m)^{1/8}} \qquad (36)$$

where d_i^{len} is measured in meters, TEC is in TEC units, frequency f is in Giga Hertz, scale height H and maximum ionization height h_m are in kilometers and elevation β in radians.

Equation (25) shows that the residual error $(\Delta s_{len})_{tr}$ depends on signal frequencies f_1, f_2, f_3 and the excess path length d_i^{len} which again depends on signal frequencies and ionospheric parameters (see Eq. 36). Substituting the excess path length expression (Eq. 36) in Eq. (25), frequency dependent terms can be separated as:

$$\begin{aligned}
(\Delta s_{len})_{tr} &= \frac{1}{C} \cdot M \cdot \left[\frac{f_1 f_3}{(f_1 - f_3)}\left(\frac{1}{f_3^2} - \frac{1}{f_1^2}\right) - \frac{f_1 f_2}{(f_1 - f_2)}\left(\frac{1}{f_2^2} - \frac{1}{f_1^2}\right) \right] \\
&= \frac{M}{f_1(f_2 - f_3)(f_1 + f_2 + f_3)} \cdot \frac{(f_2 - f_3)}{f_2 f_3} \\
&= \frac{M}{f_1 f_2 f_3 (f_1 + f_2 + f_3)}
\end{aligned} \qquad (37)$$

where,

$$M = \frac{7.5 \times 10^{-5} \cdot \exp(-2.13\beta) \cdot TEC^2}{H(h_m)^{1/8}} \qquad (38)$$

Equation (37) shows that the triple-frequency residual error $(\Delta s_{len})_{tr}$ will be positive for both the phase (see Eq. 22) and pseudo-range (see Eq. 27) combinations independently of the selection of combination frequencies. It should be mentioned here that for simplicity TEC values along signal paths are assumed to be equal in Eq. (36).

As already mentioned, higher order ionospheric corrections are derived under the assumption that the ionosphere is spherically symmetric and the vertical structure of the electron density may be described by Chapman layers. To assess the performance of the higher order corrections for real ionospheric profiles, a CHAMP based profile with $h_m = 412.3$ km, $N_m = 5.93 \times 10^{12}$ electrons/m^3 and an extreme high vertical TEC of 281.3 TECU is considered (see Figure 8). Triple-frequency residual errors $(\Delta s_{TEC})_{tr}$, $(\Delta s_{len})_{tr}$ and $(\Delta s_3)_{tr}$ have been computed by the ray tracing program and also by the correction formulas (Eqs. 31, 23; 36, 25 and 34, 24, respectively). Both ray tracing and correction results are compared in Figure 9.

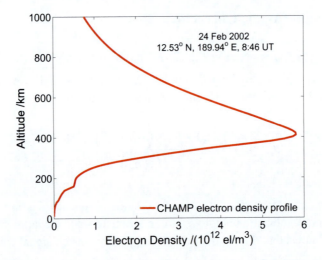

Figure 8. A CHAMP based electron density profile with h_m = 412.3 km, N_m = 4.93x10^{12} el/m^3 and vertical TEC = 281.3 TECU.

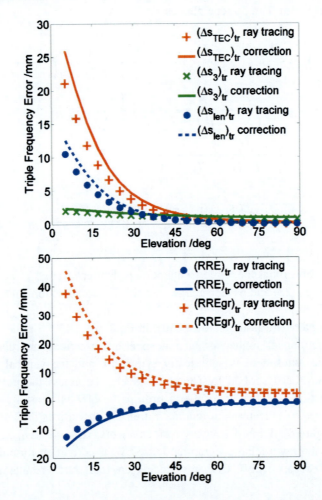

Figure 9. Validation of higher order ionospheric correction formulas.

In case of $(\Delta s_{TEC})_{tr}$ computation, the maximum remaining error, i.e., ray tracing result minus (–) correction result, is found to be about 5 mm at 5° elevation angle, whereas the maximum $(\Delta s_{TEC})_{tr}$ without correction (i.e., ray tracing result) is about 21 mm. The remaining error can be termed as error with correction and the ray tracing result as error without correction. In case of $(\Delta s_{len})_{tr}$ computation, the maximum error with correction is found to be about 2 mm at 5° elevation angle, whereas the maximum error without correction (i.e., ray tracing result) is about 13 mm. For the third order residual term $(\Delta s_3)_{tr}$, these numbers are about 0.4 mm and 2 mm, respectively. The $(RRE)_{tr}$ and $(RREgr)_{tr}$ can be corrected within accuracies of about 3 and 8 mm while without correction they can reach up to 13 and 38 mm, respectively (see Figure 9).

Statistical estimates of various higher order ionospheric effects presented here are based on CHAMP electron density profiles obtained at 2002 and 2006 which correspond to high and low solar activity conditions, respectively. For solar activity conditions higher than the referred year 2002, the statistical estimates will be higher.

Quadruple-Frequency Combination

The Galileo system will allow four frequencies E2-L1-E1, E5a, E5b and E6 (1575.42, 1176.45, 1207.14 and 1278.75 MHz, respectively) to be used in code and phase combinations for higher order ionospheric corrections. In the following, the remaining higher order ionospheric errors in the quadruple-frequency combination will be derived.

Combining Eqs. (20) and (21), the second order ionospheric term will be eliminated and we will obtain a triple-frequency (f_1, f_2, f_3) carrier-phase combination. Similarly, for f_1, f_2, f_4 frequencies, another triple-frequency combination can be obtained. Combining both triple-frequency combinations, the third order ionospheric term will be removed and we will obtain a quadruple-frequency (f_1, f_2, f_3, f_4) combination.

$$\frac{1}{H}\left[F\times\Phi\left(f_1,f_2,f_3\right)-G\times\Phi\left(f_1,f_2,f_4\right)\right]=\rho+\left(\Delta s_{len}\right)_{qd}+\left(\Delta s_{TEC}\right)_{qd} \qquad (39)$$

in which

$$\begin{aligned}
\Phi\left(f_1,f_2,f_3\right)&=\left[A\left(\Phi_1 f_1^2-\Phi_2 f_2^2\right)-B\left(\Phi_1 f_1^2-\Phi_3 f_3^2\right)\right]\\
\Phi\left(f_1,f_2,f_4\right)&=\left[A\left(\Phi_1 f_1^2-\Phi_2 f_2^2\right)-D\left(\Phi_1 f_1^2-\Phi_4 f_4^2\right)\right]
\end{aligned} \qquad (40)$$

$$\begin{aligned}
\left(\Delta s_{len}\right)_{qd}&=\frac{F}{H}\left[A\left(f_1^2 d_1^2-f_2^2 d_2^2\right)-B\left(f_1^2 d_1^{len}-f_3^2 d_3^{len}\right)\right]\\
&\quad-\frac{G}{H}\left[A\left(f_1^2 d_1^{len}-f_2^2 d_2^{len}\right)-D\left(f_1^2 d_1^{len}-f_4^2 d_4^{len}\right)\right]
\end{aligned} \qquad (41)$$

$$\begin{aligned}
\left(\Delta s_{TEC}\right)_{qd}&=\frac{F}{H}\left[A\left(\Delta TEC_2-\Delta TEC_1\right)-B\left(\Delta TEC_3-\Delta TEC_1\right)\right]\\
&\quad-\frac{G}{H}\left[A\left(\Delta TEC_2-\Delta TEC_1\right)-D\left(\Delta TEC_4-\Delta TEC_1\right)\right]
\end{aligned} \qquad (42)$$

$$D = \frac{f_1 f_4}{f_1 - f_4}$$

$$E = f_1\left(f_2 - f_4\right)\left(f_1 + f_2 + f_4\right)$$

$$F = \frac{f_2 f_3}{f_2 - f_3} \tag{43}$$

$$G = \frac{f_2 f_4}{f_2 - f_4}$$

$$H = f_1 f_2 \left(f_3 - f_4\right)\left(f_1 + f_2 + f_3 + f_4\right)$$

where $(\Delta s_{TEC})_{qd}$ is the residual error due to TEC difference along signal paths and $(\Delta s_{len})_{qd}$ is the residual error due to excess path length in the quadruple-frequency combination. Equation (39) shows that, there is no second or third order term in the combination.

Substituting A, B, D, F, G, H values and the ΔTEC_i expression (Eq. 31) in Eq. (42), it has been found that the two terms on the right hand side of Eq. (42) are equal and can be written as:

$$\frac{F}{H}\left[A\left(\Delta TEC_2 - \Delta TEC_1\right) - B\left(\Delta TEC_3 - \Delta TEC_1\right)\right]$$

$$= \frac{G}{H}\left[A\left(\Delta TEC_2 - \Delta TEC_1\right) - D\left(\Delta TEC_4 - \Delta TEC_1\right)\right] \tag{44}$$

$$= -\frac{J}{f_1 f_2 \left(f_3 - f_4\right)\left(f_1 + f_2 + f_3 + f_4\right)}$$

where J (Eq. 33) is a non-dispersive term. Since the two terms on the right hand side of Eq. (42) are equal (see Eq. 44), they will be cancelled out. Therefore, the quadruple-frequency residual error $(\Delta s_{TEC})_{qd}$ will become zero for the TEC difference approximation given by Eq. (31).

Similarly, substituting A, B, D, F, G, H values and the excess path length expression (Eq. 36) in Eq. (41), it can be shown that the two terms on the right hand side of Eq. (41) are equal. They can be written as

$$\frac{F}{H}\left[A\left(f_1^2 d_1^2 - f_2^2 d_2^2\right) - B\left(f_1^2 d_1^{len} - f_3^2 d_3^{len}\right)\right]$$

$$= \frac{G}{H}\left[A\left(f_1^2 d_1^{len} - f_2^2 d_2^{len}\right) - D\left(f_1^2 d_1^{len} - f_4^2 d_4^{len}\right)\right] \tag{45}$$

$$= \frac{M}{f_1 f_2 \left(f_3 - f_4\right)\left(f_1 + f_2 + f_3 + f_4\right)}$$

where M is given by Eq. (38). Since the two terms on the right hand side of Eq. (41) are equal, they cancel out. Therefore, the quadruple-frequency residual error $(\Delta s_{len})_{qd}$ will become zero for the excess path length approximation given by Eq. (36). It should be mentioned here that for simplicity the TEC values along signal paths are assumed to be same in Eq. (36). Due to this reason, Eqs. (41) and (45) show that the quadruple-frequency residual error (Δs_{len})$_{qd}$ is zero although the ray tracing results show that (Δs_{len})$_{qd}$ values are not zero rather very small (< 0.1 mm for the E2L1E1-E5a-E5b-E6 combination) and vary with the elevation angle.

Therefore, combining four frequencies, higher order ionospheric terms can be successfully removed from the range equation, i.e., no ionospheric correction is required in the range estimation using quadruple-frequency measurements.

New Dual-Frequency Combinations

The availability of the GPS L5 signal will allow three dual-frequency combinations such as the L1-L2, L1-L5 and L2-L5. Higher order ionospheric errors in the range estimation using the L1-L2 combination have been investigated in our previous studies [Hoque and Jakowski 2007, 2008]. It will be interesting to know higher order error estimates for other two combinations in comparison with the L1-L2 combination.

The expressions for the dual-frequency higher order ionospheric terms can be derived from Eq. (20) or Eq. (21) and generalized as

$$\frac{f_i^2}{f_i^2 - f_j^2}\Phi_i - \frac{f_j^2}{f_i^2 - f_j^2}\Phi_j = \rho \underbrace{+\Delta s_{TEC} + \Delta s_2 + \Delta s_3 - \Delta s_{len}}_{RRE} \tag{46}$$

In which

$$\Delta s_{TEC} = \frac{K(\Delta TEC_j - \Delta TEC_i)}{2\left(f_i^2 - f_j^2\right)} \tag{47}$$

$$\Delta s_2 = \frac{q}{2 f_i f_j \left(f_i + f_j\right)} \tag{48}$$

$$\Delta s_3 = \frac{u}{3 f_i^2 f_j^2} \tag{49}$$

$$\Delta s_{len} = \frac{d_j^{len} f_j^2 - d_i^{len} f_i^2}{\left(f_i^2 - f_j^2\right)} \tag{50}$$

where the index i and j denote two different GNSS carrier frequencies, the terms Δs_2 and Δs_3 are referred to as the dual-frequency second and third order residual error, respectively, Δs_{len} and Δs_{TEC} are referred to as the dual-frequency residual errors due to excess path length and TEC difference along ray paths, respectively. Adding all these terms RRE is obtained which corresponds to the dual-frequency total residual range error for the phase combination (see

Eq. 46). Similarly, the dual-frequency code pseudo-range expression can be written in terms of higher order ionospheric terms as

$$\frac{f_i^2}{f_i^2 - f_j^2}\Psi_i - \frac{f_j^2}{f_i^2 - f_j^2}\Psi_j = \rho \underbrace{-\Delta s_{TEC} - 2\Delta s_2 - 3\Delta s_3 - \Delta s_{len}}_{(RRE)_{gr}} \quad (51)$$

where RREgr is the total residual range in the code combination. Different higher order ionospheric terms are computed for the L1-L2, L2-L5 and L1-L5 combinations using a large number of CHAMP based profiles plotted in Figure 1. The maximum estimates of different higher order terms have been plotted with the elevation angle in Figures 10-12.

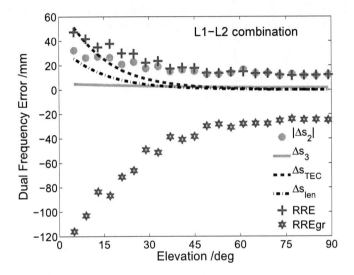

Figure 10. Maximum estimates of ionospheric errors in the L1-L2 combination.

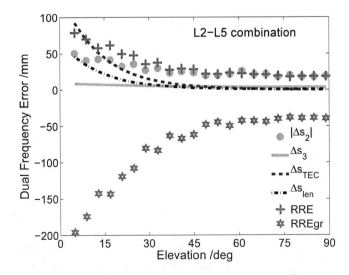

Figure 11. Maximum estimates of ionospheric errors in the L2-L5 combination.

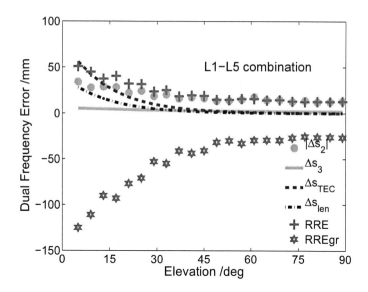

Figure 12. Maximum estimates of ionospheric errors in the L1-L5 combination.

The second order term Δs_2 can be either positive or negative depending on the direction of signal reception. Therefore, its absolute values are plotted in Figures 10-12. We see that Δs_2 plots are not smooth. The reason is that Δs_2 depends on the Earth's magnetic field and thereby varies with the azimuth angle. As already mentioned, ray tracing simulations are done at different geographic location for the elevation range 5-90°. In each case, the azimuth angle is randomly selected from the range 0-360°. Since the RRE and RREgr include the second order term, their plots show the same behaviour.

The maximum Δs_2 is estimated along North-South directions. It has been found that for a certain elevation angle and an ionospheric profile (i.e., azimuthal gradient is not considered), the maximum Δs_2 is southward (azimuth $\approx 180°$) directed when user positions are in the northern hemisphere, whereas the maximum (absolute) values are found to be in the northward (azimuth ≈ 0) direction for user positions in the southern hemisphere [Hoque and Jakowski 2006, 2007].

Comparing Figures 10, 11 and 12, we see that the ionospheric errors are minimum for the L1-L2 combination, whereas they are maximum for the L2-L5 combination. The effect of the frequency selection on the higher order ionospheric terms can be investigated by expressing them as functions of dispersive and non dispersive factors.

Substituting ΔTEC_i approximation (Eq. 31) in the dual-frequency Δs_{TEC} expression (Eq. 47), we can write

$$\Delta s_{TEC} = \frac{K}{2\left(f_i^2 - f_j^2\right)} \cdot J \cdot \left(\frac{1}{f_j^2} - \frac{1}{f_i^2}\right) = \frac{KJ}{2 f_i^2 f_j^2} \qquad (52)$$

where J is a non-dispersive term and given by Eq. (33). Equation (52) shows that Δs_{TEC} is positive independently of the frequency selection.

Substituting the excess path length approximation (Eq. 36) in the dual-frequency Δs_{len} expression (Eq. 50), we can write

$$\Delta s_{len} = \frac{1}{f_i^2 - f_j^2} \cdot M \cdot \left(\frac{1}{f_j^2} - \frac{1}{f_i^2} \right) = \frac{M}{f_i^2 f_j^2} \tag{53}$$

where M is a non-dispersive term and given by Eq. (38). Equation (53) shows that Δs_{len} is positive independently of frequency combinations.

The expressions for the dual-frequency second and third order terms are given in Eqs. (48) and (49). The quantities q and u are frequency independent terms and given by Eqs. (14) and (15), respectively. Equations (48) and (49) show that the second and third order terms are also positive independently of the frequency selection in the dual frequency combination.

Using Eq. (52) it can be shown that Δs_{TEC} is f_1^2 / f_3^2 or 1.8 times higher for the L2-L5 combination than the L1-L2 combination. However, it is f_2^2 / f_3^2 or 1.1 times higher for the L1-L5 combination than the L1-L2 combination. These relations are also valid for Δs_{len} and Δs_3 since the same dispersive term $1 / \left(f_i^2 f_j^2 \right)$ is present in their expressions (see Eqs. 53 and 49).

Again using Eq. (48) it can be shown that Δs_2 is $f_1(f_1 + f_2)/(f_3(f_2 + f_3))$ or 1.6 times higher for the L2-L5 combination and $f_2(f_1 + f_2)/(f_3(f_1 + f_3))$ or 1.1 times higher for the L1-L5 combination than the L1-L2 combination. Comparing Figures 10, 11 and 12, we see that the above findings (i.e., the effect of the frequency selection on the higher order terms) agree with simulation results.

Dividing Eq. (52) by Eq. (32), it can be shown that Δs_{TEC} is about 2.4 times higher for the L1-L2 combination than the triple-frequency L1-L2-L5 combination. Similarly, it can be shown that both Δs_3 and Δs_{len} are about 2.4 times higher for the L1-L2 combination than the triple-frequency combination. As a result, the sum of all higher order terms, i.e., *RRE* and *RREgr* are found to be more than three times higher for the L1-L2 combination than the triple-frequency L1-L2-L5 combination. Comparing Figures 6 and 7 (plots for 2002) with Figure 10, we see that ray tracing results agree with these findings.

We have found that Δs_{TEC} and Δs_3 are positive for the dual-frequency carrier-phase combination (see Eq. 46) and negative for the triple-frequency carrier-phase combination (see Eq. 22) independently of the frequency selection. The higher order term Δs_{len} is negative in the dual-frequency carrier-phase combination (see Eq. 46), whereas it is positive in the triple-frequency carrier-phase combination (Eq. 22). However, the absolute Δs_{TEC} is higher than the absolute Δs_{len} for both the dual- and triple-frequency combinations. As a result, the triple-frequency *RRE* is found to be negative, i.e., the true range ρ is larger than the determined range, whereas the dual-frequency *RRE* is found to be positive, i.e., the true range is smaller than the determined range. Similarly, it can be shown that *RREgr* is positive for the triple-frequency combination and negative for the dual-frequency combination. These relations are independently of the frequency selection.

Impact of Ionosphere Free Combination

As already described, the first order ionospheric term can be eliminated by dual-frequency measurements. Similarly, triple-frequency measurements eliminate the first and second order ionospheric terms, and quadruple-frequency measurements remove the first, second and third order ionospheric terms. However, such multiple frequency combinations lead to amplification of all uncorrelated errors or noises (multipath and noise). The noise amplification factor will be computed for different ionosphere free combinations in the following.

The following formula gives the standard deviation of the noise in the first order ionosphere free (i.e., dual-frequency) observables (Eq. 46, 51).

$$\sigma_{dual-freq} = \frac{\sqrt{f_1^4 \sigma_1^2 + f_2^4 \sigma_2^2}}{f_1^2 - f_2^2} \tag{54}$$

where σ_1 and σ_2 are the standard deviation of noise at f_1 and f_2 frequencies, respectively. For simplicity, we can assume that the measurement noise is the same on each of the two frequencies, i.e., $\sigma_1 = \sigma_2$. Using Eq. (54), it can be shown that the phase or code noise will be amplified by a factor of 2.98 for the L1-L2 combination, whereas for the L1-L5 and L2-L3 combinations amplification factors are 2.59 and 16.64, respectively. The amplification factor is inversely proportional to the separation of frequencies. For the L1-L5 combination the frequency separation is relatively large and therefore the amplification factor is the smallest.

The triple-frequency combination Eq. (22) can be simplified as

$$\frac{\Phi_1 f_1^3 \left(f_2 - f_3\right) + \Phi_2 f_2^3 \left(f_3 - f_1\right) + \Phi_3 f_3^3 \left(f_1 - f_2\right)}{\left(f_1 + f_2 + f_3\right)\left(f_1 - f_2\right)\left(f_2 - f_3\right)\left(f_1 - f_3\right)} = \rho + \left(RRE\right)_{tr} \tag{55}$$

Equation (55) reveals that the standard deviation of the noise in the triple-frequency combination can be written as

$$\sigma_{triple-freq} = \frac{\sqrt{\sigma_1^2 f_1^6 \left(f_2 - f_3\right)^2 + \sigma_2^2 f_2^6 \left(f_3 - f_1\right)^2 + \sigma_3^2 f_3^6 \left(f_1 - f_2\right)^2}}{\left(f_1 + f_2 + f_3\right)\left(f_1 - f_2\right)\left(f_2 - f_3\right)\left(f_1 - f_3\right)} \tag{56}$$

where σ_1, σ_2 and σ_3 are the standard deviation of the noise at f_1, f_2 and f_3 frequencies, respectively. For simplicity, we can again assume that $\sigma_1 = \sigma_2 = \sigma_3$. Using Eq. (56), it can be shown that the noise will be amplified by a factor of 33.7 in the combined L1-L2-L5 signal.

Similarly, it can be shown that the noise estimation for the quadruple-frequency combination will be

$$\sigma_{quadruple-freq} = \frac{\sqrt{\sigma_{11}^2 + \sigma_{22}^2 + \sigma_{33}^2 + \sigma_{44}^2}}{b} \tag{57}$$

$$\left. \begin{aligned} \sigma_{11} &= \sigma_1 f_1^4 \left(f_2 - f_3\right)\left(f_3 - f_4\right)\left(f_2 - f_4\right) \\ \sigma_{22} &= \sigma_2 f_2^4 \left(f_1 - f_3\right)\left(f_3 - f_4\right)\left(f_1 - f_4\right) \\ \sigma_{33} &= \sigma_3 f_3^4 \left(f_1 - f_2\right)\left(f_1 - f_4\right)\left(f_2 - f_4\right) \\ \sigma_{44} &= \sigma_4 f_4^4 \left(f_1 - f_2\right)\left(f_1 - f_3\right)\left(f_2 - f_3\right) \end{aligned} \right\} \tag{58}$$

$$b = \left(f_1 + f_2 + f_3 + f_4\right)\left(f_1 - f_2\right)\left(f_1 - f_3\right)\left(f_1 - f_4\right)\left(f_3 - f_4\right)\left(f_2 - f_4\right)\left(f_2 - f_3\right) \qquad (59)$$

Using Eq. (57), it can be shown that the noise will be amplified by a factor of about 626.13 in the quadruple-frequency combination E2L1E1-E5a-E5b-E6. Thus the quadruple-frequency combination although eliminates the higher order ionospheric errors successfully, it amplifies the noise contribution by a factor which is about two order higher than a dual-frequency factor. Therefore, the quadruple-frequency measurements are barely pragmatic. However, if the frequency separations are large (e.g., combinations between 4-8 GHz C band and 1-2 GHz L band frequencies), the amplification factor will be small in the quadruple-frequency measurements. In such cases, measurements from four frequencies may be useful.

For a signal-to-noise ratio of 35 dB-Hz, the one sigma standard deviation of the measurement noise (only thermal noise contribution is considered) is about 4 mm for the GPS-L1 phase, whereas it may exceed the 60 cm level for the GPS-P(Y) code [Irsigler et al. 2004]. Considering the same measurement noise for the L2 and L5 phases as the L1 phase, it can be shown that the phase noise is about 1.2 cm and 1 cm for the dual-frequency L1-L2 and L1-L5 combinations, respectively, and about 6.7 cm for the L2-L5 combination. Similarly, considering the measurement noise for each of the three codes as 60 cm, it can be shown that the code noise is about 1.8 m and 1.6 m for the L1-L2 and L1-L5 combinations, respectively, and about 10 m for the L2-L5 combination. However, the first order ionospheric term (a few meters to tens of meters at zenith) will be eliminated from the dual-frequency measurements, which is much more than the induced noise.

As already shown, the measurement (phase or code) noise will be amplified by a factor of 33.7 in the triple-frequency combination. Therefore, the triple-frequency measurement noise is up to 13.5 cm and 20.2 m in the phase and code combinations, respectively. The impact of ionosphere free combinations on the measurement noise has been summarized in the following Table 1.

Table 1. Impact of ionosphere free combinations on measurement noise.

Combination	Noise amplification factor	Phase noise	Code noise
L1-L2	2.98	1.2 cm	1.8 m
L1-L5	2.59	1 cm	1.6 m
L2-L5	16.64	6.7 cm	10 m
L1-L2-L5	33.7	13.5 cm	20.2 m
E2L1E1-E5a-E5b-E6	626.13		

Figures 10, 11, and 12 show that the dual-frequency RRE and $(RRE)_{gr}$ can reach up to 5 cm and 13 cm, respectively, in the L1-L2 and L1-L5 combinations, whereas for the L2-L5 combination these numbers are about 9 cm and 20 cm, respectively. Comparing dual- and triple-frequency residual errors (compare Figures 10, 11, and 12 with Figures 6 and 7), we see that RRE is reduced from about 5/9 cm to about 1.3 cm and $(RRE)_{gr}$ from about 13/20 cm to about 4 cm for the triple-frequency measurements.

We see that the increase of the measurement noise (from about 1/7 cm to 13.5 cm for phases and about 2/10 m to 20 m for codes) is higher than the reduction of the ionospheric errors in the triple-frequency combinations. Thus the uncertainty in the noise exceeds the uncertainty in the ionospheric errors for the triple-frequency measurements.

Conclusion

A large number of ionospheric electron density profiles (about 50,000) obtained from CHAMP-GPS radio occultation measurements has been used to calculate different higher order ionospheric errors in the GNSS triple-frequency measurements. The triple-frequency measurements eliminate the first and second order ionospheric terms; but the third order ionospheric term and errors due to ray path bending are not fully compensated in this approach. We have found that at $5°$ elevation angle the residual range errors reach the 1 cm and 4 cm levels for the phase and code pseudo-range measurements, respectively, at high level of total electron content.

It has been found that the triple-frequency ionospheric errors are about one third of the dual-frequency ionospheric errors. However, the quantitative reduction of the ionospheric errors is less than the increase of the measurement noise in the triple-frequency measurements. Therefore, the triple-frequency measurements for higher order ionospheric correction are practically not useful. Instead the dual-frequency measurements and correction of their higher order terms are practically useful, since the measurement noise will not be further increased by the correction.

The Galileo system will provide four frequencies to be used in code and phase measurements. It has been found that combining four frequencies, the ionospheric effects can be cancelled out successfully from range estimations. However, the measurement noise will be amplified by a factor of order two, which makes the quadruple-frequency measurements barely pragmatic.

References

Appleton E. V. (1932), Wireless studies of the ionosphere, *J. Instn. Elect. Engrs.* **71**, 642-50

Bassiri S. and G. A. Hajj (1993), Higher-order ionospheric effects on the global positioning system observables and means of modeling them, *manuscripta geodaetica*, **18**(6), 280-289.

Brunner F. K. and M. Gu (1991), An improved model for the dual frequency ionospheric correction of GPS observations, *manuscripta geodaetica*, **16**(3), 205-214.

Booker H. G. (1949), Application of the magnetoionic theory to radio waves incident obliquely upon a horizontally stratified ionosphere, *J. Geophys. Res.*, **54**, 243-74.

Budden K. G. (Ed.) (1985), *The Propagation of Radio Waves: the theory of radio waves of low power in the ionosphere and magnetosphere*, Cambridge University Press, Cambridge.

Chen J., P. L. Dyson and J. A. Bennet (1990), Automatic fitting of Quasi-Parabolic Segments to Ionospheric profiles with application to Ground Range Estimation for single-station location, *Journals of Atmospherics and Terrestrial Physics*, **52**(4), 277.

Davies K. (Ed.) (1990), *Ionospheric Radio*, Peter Peregrinus Ltd, London.

Hartmann G. K. and R. Leitinger (1984), Range errors due to ionospheric and tropospheric effects for signal frequencies above 100 MHz, *Bull. Geod*, **58**(2), 109-136.

Haselgrove J. (1963), The Hamiltonian ray path equation, *J. Atmos. Terr. Phys.*, **25**, 397-399.

Hartree D. R. (1931), The propagation of electromagnetic waves in a refracting medium in a magnetic field, *Proc. Camb. Phil. Soc.* **27**, 143-62

Hoque M. M. and N. Jakowski (2006), Higher-order ionospheric effects in precise GNSS positioning, *Journal of Geodesy*, **81**(4), 259-268, DOI 10.1007/s00190-006-0106-0.

Hoque M. M. and N. Jakowski (2007), Mitigation of higher order ionospheric effects on GNSS users in Europe, *GPS Solutions*, **12**(2), DOI 10.1007/s10291-007-0069-5

Hoque M. M. and N. Jakowski (2008), Estimate of higher order ionospheric errors in GNSS positioning, *Radio Sci.*, **43**, DOI 10.1029/2007RS003817

Irsigler M., G. W. Hein, A. Schmitz-Peiffer (2004) Use of C-Band frequencies for satellite navigation: benefits and drawbacks, *GPS Solution*, **8**:119-139, doi: 10.1007/s10291-004-0098-2

Jakowski N., F. Porsch and G. Mayer (1994), Ionosphere-Induced-Ray-Path Bending Effects in Precise Satellite Positioning Systems, *Zeitschrift für Satellitengestützte Positionierung, Navigation und Kommunikation, SPN* **1/94**, 6-13.

Jakowski N., A. Wehrenpfennig, S. Heise, C. Reigber, H. Lühr, L. Grunwaldt, T. K. Meehan (2002), GPS radio occultation measurements of the ionosphere from CHAMP: Early results, *Geophys. Res. Lett,* **29**(10), 1457, DOI 10.1029/2001 GL014364

Jakowski N. (2005), Ionospheric GPS Radio Occultation measurements on board CHAMP, *GPS Solutions*, **9**, 88–95, doi: 10.1007/s10291-005-0137-7

Jones R. M., J. J. Stephenson (1975), *A Versatile Three-Dimensional Ray Tracing Computer Program for radio waves in the ionosphere*, Technical Report OTR-75-76, U. S. Department of Commerce, Office of Telecommunications, Boulder, Colorado

Kelso J. M. (1975), Radio ray propagation in the ionosphere, McGraw-Hill, New York, ISBN 111434172X

Klobuchar J. A. (1996), Ionospheric Effects on GPS, *In Global Positioning System: Theory and Applications*, Vol I, edited by Parkinson B. W. and J. J. Spilker, pp. 485-515, *American Institute of Aeronautics and Astronautics*, ISBN 156347106X.

Lassen H. (1927), Über den Einfluss des Erdmagnetfeldes auf die Fortpflanzung der Elektrischen Wellen der drahtlosen Telegraphie in der Atmosphäre, *Elektrische Nachrichten Technik*, **4**, 324-34

Mandea M. and S. Macmillan (2000), International Geomagnetic Reference Field- the eighth generation, *Earth Planets Space*, **52**(12), 1119–1124.

Nelder J. A., R. Mead (1965), A Simplex Method for Function Minimization, *Computer J*, **7**, 308-313.

Reigber Ch., H. Lühr, P. Schwintzer (2002), CHAMP mission status, *Adv. Space Res.*, **30**, 129-134.

Rishbeth H. and O. K. Garriott (Eds.) (1969), *Introduction to ionospheric physics*, Academic Press, New York.

Tucker A. J., B. M. Fannin (1968), Analysis of ionospheric contributions to the Doppler shift of CW signals from artificial earth satellites, *J. Geophys. Res.,* **73**, 4325-4334.

In: Global Positioning Systems
Editors: V. Asphaug and E. Sørensen, pp. 29-41

ISBN: 978-1-60741-012-6
© 2010 Nova Science Publishers, Inc.

Chapter 2

HIGHWAY GEOMETRY DETERMINATION FROM GPS DATA

María Castro[1], Luis Iglesias[2], Roberto Rodríguez-Solano[3] and José A. Sánchez[4]

[1] Departamento de Transportes, E.T.S.I.C.C.P.,
Universidad Politécnica de Madrid, Spain
[2] Departamento de Explotación de Recursos Minerales y Obras Subterráneas,
E.T.S.I.M., Universidad Politécnica de Madrid, Spain
[3] Departamento de Construcción y Vías Rurales, E.U.I.T.F.,
Universidad Politécnica de Madrid, Spain
[4] Departamento de Ingeniería Civil: Hidráulica y Energética,
E.T.S.I.C.C.P., Universidad Politécnica de Madrid, Spain

Abstract

In many situations, highway geometry is unknown. Using GPS data to reconstruct highway geometry could be the best option in a variety of circumstances. The present work analyzes the performance of several GPS receivers in this task. Three different commercial receivers have been chosen for testing and performance comparison. The three devices reach submetric accuracy. The data have been collected in kinematic mode, with the GPS antennas attached to stable magnetic bases on the top of a vehicle, and post processed by differential corrections.

Tests have been carried out on a two-lane rural highway located on Spain. Using the GPS positions, a method for highway geometry reconstruction is presented. The obtained results offer versatile solutions for multiple civil engineering applications, including the incorporation of these data in GIS environments. In this way, analysis of traffic safety and highway consistency, and other kinds of studies, could be done easily.

Introduction

GPS receivers have many uses in highway engineering. GPS devices, fitted to specialized vehicles and used in combination with Geographic Information Systems (GIS) (Quiroga

2000, Taylor et al. 2000), have been employed for studies on traffic congestion, vehicle speed (Jiang and Li 2002, Wang 2006), travel time and delay (Faghri and Hamad 2002, Pan et al. 2007), and could be used to improve crash location data (Graettinger et al. 2001). Also, GPS receivers have been used for obtaining geographic information of highways in specialized vehicles for roadway inventory (Federal U.S. Highway Administration 2000 and 2005, Transportation Research Board 2002; Ben-Arieh et al. 2004, Chang 2004). The majority of instrumented vehicles for roadway inventory use expensive positive inertial navigation units with the addition of geographic positioning systems receivers. Nevertheless, sometimes the precision of the collected data or the alignment definition by means of straight lines makes it inadequate to conduct detailed studies about the highway geometry. Some researchers have developed low cost procedures that allow highway geometry reconstruction without the need for expensive vehicles specifically equipped for making inventories (Imran et al. 2006, Castro et al. 2006, Campbell et al. 2007).

The present work analyzes the performance of several GPS receivers for reconstructing highway geometry in a reliable and economical way. Data have been collected by means of GPS receivers placed on a normal-sized passenger car. Tests have been carried out on a two-lane rural highway. Using the GPS data captured, a method for highway geometry reconstruction is presented.

Case Study

Tests have been carried out with different GPS receivers, not in real time, on a two-lane rural highway (M-607) located in Madrid (Spain). The road runs through rolling terrain and is approximately 12 km long. Each lane is 3.50 m wide, with 2.50 m shoulders. There were no urban city areas, bodies of water, trees, or other features near the M-607 highway that could affect the GPS data collection (Figure 1). The Average Annual Daily Traffic (AADT) is 15343.

GPS Devices

Devices from two different manufacturers, Leica and Trimble, were chosen for testing and performance comparison. From Trimble, the Geo XT model was chosen. From Leica, a Leica System 500, with two different receivers (SR 510 in a case and SR 530 in the other) were used. These receivers are specially designed to meet the mapping and data collection needs of a variety of industries and integrate field survey information perfectly with AutoCad, ArcView or other drafting and GIS computer systems. The three devices reach submetric accuracy.

The *Leica System 500 GPS* is composed of a terminal (TR 500), a receiver (SR 510 in one case and SR 530 in the other) and an antenna of the microstrip type with a built-in ground plane (AT 501). The terminal allows parameters and functions to be defined, by means of an alpha-numeric keyboard, and configures a record of the entries (Figure 2). This device was connected by means of a cable to the receiver; both were located in the interior of the vehicle.

Figure 1. M-607 highway.

Figure 2. Terminal (TR 500) of Leica System 500 GPS.

The antenna, connected to the receiver by means of a coaxial cable, was attached to the top of the vehicle, threaded in a standard magnetic base (Figure 3). The GPS was powered by two batteries, of the camcorder type, that guarantee, in continuous operation, an average autonomy of seven and a half hours. The logged data were stored in a 16 MB PCMCIA memory card. SR 510 receiver has 12 L1 channels with code C/A, technique of narrow correlation, code of precision and complete phase. The SR 510 receiver data recording rate is selectable from 1 to 60 sec (1 second was selected). This selection was conditioned by the data recording rate of the base station. The recommended cut-off angle is 10-25 degrees for optimal results in kinematic tests. The GPS SR 510 incorporates a choke-ring antenna that rejects multipath contamination. This GPS receiver performs extremely well under forest canopy and other difficult conditions. The nameplate mean square error of this device with post process of L1 code is 30 cm. The nominal technical specification for this device in kinematic mode after initialization is 20 mm + 2 ppm (rms).

Figure 3. Leica System 500 GPS: antenna settled on the top of a vehicle threaded in a standard magnetic base.

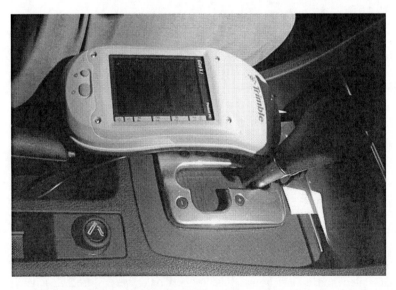

Figure 4. Terminal of Trimble Geo XT.

Figure 5. Patch Antenna and coaxial cable (Trimble Geo XT).

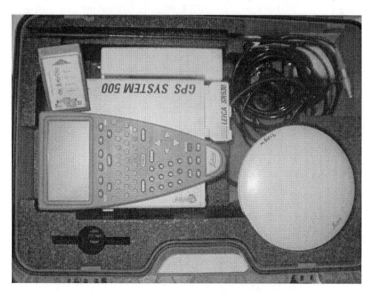

Figure 6. Leica GPS SR 530 receiver.

The *Trimble Geo XT* is composed of a handheld terminal and a Patch Antenna. The terminal has integrated a high-performance GPS/SBAS receiver and a L1 antenna (Figure 4 and 5), with 14 channels (12 L1 channels code and carrier and 2 SBAS). The terminal allows parameters and functions to be defined, by means of a touchscreen, and its operating system was Windows Mobile version 6. Data were collected using Trimble's TerraSync software. The Patch Antenna, connected to the receiver by means of a coaxial cable, was attached to the top of the vehicle. The data were stored in the internal memory of the terminal. A data recording rate of 5 second was selected. The horizontal root mean square (HRMS) accuracy for 1 sigma is less than 1 meter (submetric).

The *Leica GPS* SR 530 receiver (Figure 6) is a geodetic receiver with centimeter level accuracy. The basic technical specifications are: 12 L1 C/A narrow code, and full phase, and 12 L2 P code and phase. With the SR 530 receiver a data recording rate of 0.1 second was selected. The recommended cut-off angle is 10-25 degrees for optimal results in kinematic tests. The GPS SR 530 incorporates a choke-ring antenna that rejects multipath contamination. The baseline rms (post processing) in kinematic mode after initialization is 10 mm + 1 ppm.

Data Collection

Data collection was performed using three schemes: using (i) Leica System 500 GPS with SR 510 receiver, (ii) Trimble GEO XT, and (iii) Leica System 500 GPS with SR 530 receiver.

With *Leica SR 510 receiver* data were registered with the minimum interval of 1 second. The measurement technique applied was kinematic mode after a static known point initialization (KPI). C/A code and full-wave carrier phase were processed with no loss of lock. The PDOP value was between 6.9 and 2.1 with an average value of 4.8. The number of satellites available during observation was between 8 and 9. The mask angle was set to 20 degrees maintained during the kinematic observation to avoid the noise induced by low elevation satellite signals.

Figure 7. Vertex (LA OLLERA) of the Spanish geodetic network with Leica GSP System 500 (SR 530 receiver).

With *Trimble Geo XT* data were registered with a minimum interval of 5 seconds. The measurement technique applied was kinematic mode after a static known point initialization (KPI). C/A code and L1 carrier phase were processed with no loss of lock. The PDOP value was between 1.7 and 3.1, with an average value of 2.4. The number of satellites available during observation was between 6 and 9. The mask angle was set to 15 degrees maintained during the kinematic observation.

With *Leica SR 530 receiver* data were registered with the minimum interval of 0.1 seconds. The measurement technique applied was kinematic mode after a static known point initialization (KPI). C/A code and full-wave carrier phase were processed with loss of lock. Two GPS units were used, one as base station and another as rover. The base station GPS was placed above a vertex of the Spanish geodetic network (REGENTE). The PDOP value was between 4.8 and 9.0 with an average value of 5.5. The number of satellites available during observation was between 6 and 9. The mask angle was set to 20 degrees maintained during the kinematic observation. Real Time Kinematic (RTK) techniques were impractical because the communication link between the reference receiver and the mobile receiver was not possible.

In all cases, during data logging, the vehicle was maintained at an approximate speed of 80 km/h. The speed of driving is very conditioned by the highway alignment and its traffic (AADT = 15343) because the data logging vehicle must not be driven at a speed far below the average speed of the route in order to avoid becoming an obstacle for other vehicles. So, points recorded with the Leica GPS SR 510 receiver are, approximately, 20 m apart from one other; with Trimble Geo XT, 100 m; and with Leica GPS SR 530 receiver, 2 m.

Data Post-Processing

For both *Leica SR 510 GPS receiver and Trimble GEO XT* cases, differential correction processes have been made using a base station located in the Escuela Técnica Superior de Ingenieros en Topografía, Geodesia and Cartografía of the Universidad Politécnica of Madrid. This base station is denominated GPS MERC and provides continuous GPS data since 1998. It is equipped with a Trimble CORS SSi 4000 receiver with 12 channels for the L1 carrier phase and another 12 for the L2 carrier. It collects C/A code on the L1 and P code on the L2, even if it is encrypted. It has a L1/L2 geodetic antenna of the 'choke ring' type with a Dorne Margolin element. The data storage and RINEX format conversion is done with Trimble Reference Station software. The data are recorded in various formats with an interval of at least 1 second (1Hz). This station is materialized by a pillar one meter high topped by a metal plate with forced centering sockets in which the antenna goes directly threaded (Figure 8). The station coordinates refer to a point called the reference point of the station which coincides with this plate. The station coordinates for GPS MERC are presented in Table 1. The approximate distance between the base station and the studied highways is less than 38 km.

For *Leica SR 530 GPS receiver* the base station used for doing the differential correction process was located in a vertex of the Spanish geodetic network denominated LA OLLERA. The vertex is materialized by a pillar 1.19 m tall and 0.29 m diameter, located on a bucket with 1 m sides (Figure 7). The pillar is topped by a metal plate with forced centering sockets. The antenna of one of the two Leica System 500 GPS with SR 530 receiver used for data capture, were installed using forced centering sockets. The station coordinates are referred to the plate located in the top of the pillar. The station coordinates for LA OLLERA vertex are presented in Table 1. The approximate distance between the base station and the studied highways is less than 17 km.

Figure 8. Base station GPS MERC.

Table 1. Characteristics of base stations

Station	Latitude	Longitude	Ellipsoidal Altitude	X UTM - 30N Zone	Y UTM - 30N Zone	Orthométric Altitude
Merc	ETRS89 Geodetic System					
	40° 23' 21,70421" N	3° 37' 48,08926" W	727,030	446527,794	4471164,584	
	ED-50 Geodetic System					
	40° 23' 25,9600" N	3° 37' 43,3827" W		446637,271	4471372,159	676,119
La Ollera	ETRS89 Geodetic System					
	40° 40' 01,43942" N	3° 43' 43,64353" W	941,805	438400,025	4505054,651	
	ED-50 Geodetic System					
	40° 40' 05,66240" N	3° 43' 38,92000" W		438509,220	4502262,000	889,200

Orthometric altitude referred to the mean sea level in Alicante (Spain) (NMMA).

With *Leica System 500 GPS with SR 510 receiver*, Ski –Pro L1 software by Leica was used for differential post-processing with broadcast ephemeredes. With *Trimble Geo XT* GPS

Pathfinder Office software by Trimble was used for differential prost-processing. With *Leica System 500 GPS with SR 530 receiver* GIS DataPro and Leica Geo Office software was used.

Determination of the Roadway Centerline

Data collected with GPS devices do not correspond with the roadway centerline because the vehicle that logs the data follows a path centered on its lane and does not go over the centerline (the antenna is placed centered on the top of the vehicle). Data from both lanes have been taken, so there are two paths. Nevertheless, it can be assumed that both paths are symmetrical with respect to the centerline. From this assumption, a procedure that allows obtaining points located, approximately, on the roadway centerline has been developed (Castro et al. 2006).

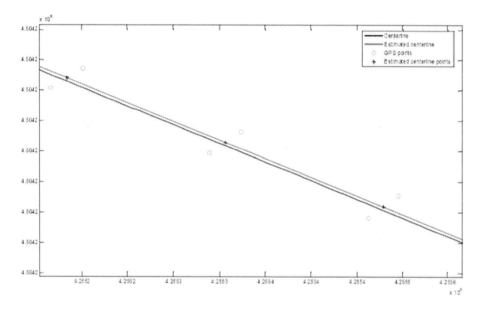

Figure 9. Detail of the estimation of centerline from data taken with GPS (Leica SR 510 receiver).

This procedure searches, for each point of a path, the nearest point of the other path. The midpoint between these two points will be a point of the centerline (Figure 9). The procedure used to find the nearest point of the other path consisted in computing the geometric distance between each point of a path and all the points of the other path, the point at minimum distance will be the nearest point. A byproduct of this procedure is an estimation of the "width" of a lane of the highway in each estimated point of the centerline. The information provided by these lane "widths" allows the evaluation of the error made in the estimation of points of the centerline. This procedure of estimation has been implemented in Matlab.

The proposed procedure uses the GPS data logged from both lanes of the highway. So, an estimation of points of the roadway centerline closer to reality than in the procedure proposed by Ben-Arieh et al. (2004), which uses only data from one of the paths chosen arbitrarily, is obtained.

Once the points of the centerline have been estimated several possibilities of geometry determination appear. One possibility, proposed by Imran et al. (2006) and Campbell et al. (2007), is to estimate the location and parameters of tangents and circular curves. It should be noted that, in the transition curves (spirals), these authors do not determine their parameters. Another possibility, proposed by Ben-Arieh et al. (2004) and Castro et al. (2006), is using splines to define the centerline. Ben-Arieh et al. (2004) use B-splines while Castro et al. (2006) use smoothing splines. As the coordinates of the estimated points of the centerline have errors, both procedures do some smoothing of these data.

A spline is a piecewise polynomial interpolating function that crosses all known points of the line interpolated. The degree, n, of the spline is the degree of the interpolating polynomials. In the known points, the spline is continuous and its $(n-1)$ derivatives are also continuous. The known points are called crossing points or knots of the spline. As splines with odd degrees have graphical representations smoother than those with even degrees, and, as using polynomials of low degree is simpler, cubical splines are the most used. Therefore, they are piecewise polynomial functions, continuous and with first and second derivatives continuous as well.

Due to the inherent errors of the described procedure of estimating points of the centerline, it would be better to use smoothing than interpolating splines. Smoothing splines are defined with a variational procedure as the minimum of the following functional (Castro et al. 2006):

$$\rho \sum_{j=1}^{n} w(j)\left|y_j - f(x_j)\right|^2 + \int \lambda(t)\left|D^2 f(t)\right|^2 dt \qquad (1)$$

where, n is the number of values of x, and the integral extends over the smallest value of the interval of x. The default value of the weighting vector $w(j)$ is one. The default value of the roughness function λ is also one. D^2 represents the second derivative of the function f(t), and ρ is the smoothing parameter. This parameter is defined indirectly by means of a tolerance *tol*. So, the term to which it multiplies has the value *tol*. Therefore:

$$tol = \sum_{j=1}^{n} w(j)\left|y_j - f(x_j)\right|^2 \qquad (2)$$

This tolerance could be defined from an estimation of the error committed in the determination of the points of the centerline. The procedure for estimation of points of the centerline also provides a distribution of "widths" of a lane of the highway (Castro et al. 2006). If the points were always faced and equidistant from the highway centerline, this distribution of "widths" would be a constant. As this is not the case, the standard deviation of this distribution is a measurement of the error in the determination of the width of a lane of the highway as the average of the distribution of widths. Therefore, half of this value is a measurement of the error obtained in the estimation of points of the centerline. So, the value of the tolerance *tol* used is half this standard deviation.

Smoothing splines could be defined in Cartesian and in parametric coordinates. In highways with sharp curves, more than one Y coordinate could exist for the same coordinate

X, and a similar problem could arise in designs sensibly parallel to the Y axis. Thus, a definition of smoothing splines in parametric coordinates is recommended.

Comparative Study

Results obtained (Figure 10) using the *Trimble Geo XT* GPS were very good (PDOP between 1.7 and 3.1), but, as it was configured to collect data every 5 s only, points obtained were far apart (100 m). As a result, the maximum distance between the estimated centerline and the centerline defined in the Project is about 2 m.

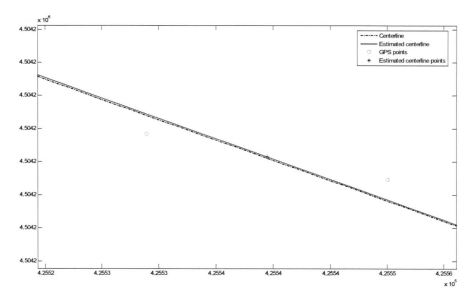

Figure 10. Detail of the estimation of centerline from data taken with Trimble Geo XT GPS.

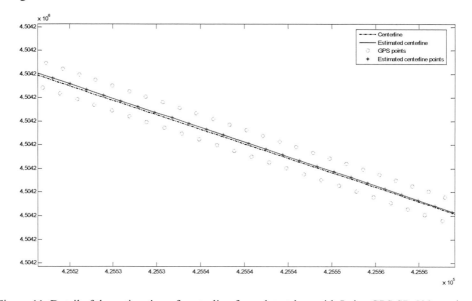

Figure 11. Detail of the estimation of centerline from data taken with Leica GPS SR 530 receiver.

The *Leica GPS SR 510 receiver* was configured to collecting data every second. Thus, points obtained were approximately 20 m apart from each other (Figure 9). The maximum difference between the estimated centerline and the centerline defined in the Project is about 1 m.

Using the *Leica GPS SR 530 receiver*, data were collected every 0.1 s and, as a consequence, points obtained were circa 2 m apart from each other. However, results obtained were not better (Figure 11) than in previous cases. In addition, it must be taken into account that for collecting data every 0.1 s two stations and three people are needed. When data were collected every second, or every several seconds, only one station and two people are needed because the differential correction could be done using a fixed nearby base station. Thus, this procedure (two stations) is more complicated and costly than the other and do not provides better results.

Conclusion

Three different receivers have been chosen for testing and performance comparison on a two-lane rural highway located in Spain. The three devices reach submetric accuracy. The data have been collected by attaching the GPS antennas to the top of a vehicle with stable magnetic bases.

Using the GPS positions, a method for highway geometry reconstruction is presented. The results obtained offered versatile solutions for multiple civil engineering applications, including the incorporation of these data in GIS environments. In this way, analysis of traffic safety and highway consistency, and other kinds of studies, could be done easily.

The analysis in the comparative study indicates that the additional costs needed to obtain positions every 0.1 s cannot be justified because these results were not better than those obtained from positions registered every second.

Acknowledgements

The authors thank the *Consejería de Transportes e Infraestructuras* of the *Comunidad de Madrid*, for its collaboration in this research.

References

Ben-Arieh, D., Chang, S., Rys, M., and Zhang, G., (2004). Geometric modeling of highways using global positioning system data and B-spline approximation. *Journal of Transportation Engineering*, **130** (5), 632-636.

Campbell, F., Ogle, J. H., Rhoades, J., Sarasua, W. A., and Chowdhury, M. A. (2007). Evaluation of Methods for Determining Horizontal Road Design Data Using an Instrumented Vehicle Equipped with Satellite Differentially Corrected GPS. *Transportation Research Board 86th Annual Meeting*. Transportation Research Board.

Castro, M., Iglesias, L., Rodríguez-Solano, R., and Sánchez, J. A. (2006). Geometric modelling of highways using global positioning system (GPS) data and spline approximation. *Transportation Research. Part C* **14** (4), 233-243.

Chang, S. I., 2004. *Global Positioning System data integration and development of a three-dimensional spatial model of the Kansas highway network.* FHWA-KS-03-7, Topeka, KS.

Faghri, A. and Hamad, K. (2002). Travel time, speed, and delay analysis using an integrated GIS/GPS system. *Canadian Journal of Civil Engineering*, **29** (2), 325–328.

Federal Highway Administration (2005). *High accuracy-nationwide differential global positioning system test and analysis: Phase II report.* FHWA-HRT-05-034, McLean, VA.

Federal Highway Administration (2000). *An investigation of the use of Global Positioning System (GPS) technology and its augmentations within state and local transportation departments.* FHWA-RD-00-093, McLean, VA.

Graettinger, A. J., Rushing, T. W., and McFadden, J. (2001). Evaluation of inexpensive global positioning system units to improve crash location data. *Transportation Research Record,* **1746**, 94-101.

Imran, M., Hassan, Y., and Patterson, D. (2006). GPS-GIS-based procedure for tracking vehicle path on horizontal alignments. *Computer-Aided Civil and Infrastructure Engineering*, **21** (5), 383-394.

Jiang, Y., and S. Li. (2002). Measuring and Analyzing Vehicle Position and Speed Data at Work Zones Using Global Positioning Systems. *ITE Journal,* **72** (3), 48-53.

Pan, C., Lu, J., Wang, D., and Ran, B. (2007). Data Collection Based on Global Positioning System for Travel Time and Delay for Arterial Roadway Network. *Transportation Research Record*, **2024**, 35-43.

Quiroga C. A. (2000). Performance measures and data requirements for congestion management systems. *Transportation Research Part C: Emerging Technologies*, **8** (1-6), 287-306.

Taylor, M. A., Woolley, J. E., and Zito, R., 2000. Integration of the global positioning system and geographical information systems for traffic congestion studies. *Transportation Research Part C: Emerging Technologies*, **8** (1-6), 257-285.

Transportation Research Board (2002). *Collecting, Processing, and Integrating GPS Data into GIS. A Synthesis of Highway Practice.* NCHRP Synthesis 301, Washington, DC.

In: Global Positioning Systems
Editors: V. Asphaug and E. Sørensen, pp. 43-55

ISBN: 978-1-60741-012-6
© 2010 Nova Science Publishers, Inc.

Chapter 3

HOW LOCATION PERFORMANCE INDEXES OF GPS RADIO COLLAR REFLECT LOCATION ERROR IN MOUNT FUJI, CENTRAL JAPAN

Zhaowen Jiang[1,][a], Seiki Takatsuki[2,b], Masahiko Kitahara[1] and Mikio Sugita[3]*

[1] Laboratory of Animal Ecology,
Yamanashi Institute of Environmental Sciences, Yamanashi, Japan
[2] The University Museum, The University of Tokyo, Tokyo, Japan
[3] Laboratory of Land Planning and Resource Use,
Yamanashi Institute of Environmental Sciences, Yamanashi, Japan

Abstract

We assessed the effects of habitat features with varying canopy closures, tree density, tree basal area, and topography on location performance of GPS radio collars (GPS3300, Lotek Engineering Inc., Ontario, Canada) around Mount Fuji, a single peak surrounded by wide and flat areas. We analyzed the differences of performance indexes (n = 1361) recorded by the GPS radio collars among different location status (unsuccessful fix, successful 2-dimensional (D), and 3-D location) and the relationships between indexes related location quality by successful location data. There were differences in all indexes (available satellite, location time, and location data dimension) among or between location status but not in position dilution of precision (PDOP). Available satellite number increased from unsuccessful to 2-D and to 3-D location. In contrast to available satellite, location time decreased. The differences between location indexes reflect location error and showed that 2-D location was greater than 3-D; however, the difference between 2- and 3-D was not reflected PDOP. Correlation analyses showed that available satellite number affected location time (r = -0.390),

---

[*] E-mail address: jiang@wmo.co.jp. Tele. +81-427-98-7545, Fax +81-427-98-7565. Corresponding author: Zhaowen Jiang, Wildlife Conservation and Management Office, Oyamagaoka 1-10-13, Machida, Tokyo 194-0215, Japan.
[a] Present address: Zhaowen Jiang, Wildlife Conservation and Management Office, Oyamagaoka 1-10-13, Machida, Tokyo 194-0215, Japan.
[b] Seiki Takatsuki, The Laboratory of Wildlife Biology, Azabu University, Fuchinobe 1-17-71, Sagamihara, Kanagawa 229-8501, Japan.

location data dimension (r = 0.639), PDOP (r = -0.362), and, finally, location error (r = -0.282). The location time, PDOP, and location error decreased with the increasing available satellite number. A higher satellite number yielded greater 3-D location. Meanwhile, small PDOP reflected less location error (r = 0.407) and shorter location time (r = 0.088). However, PDOP did not show a relation with the proportion of location data dimension (r = 0.042). Further, longer location time correlated with greater location error (r = 0.196) and less proportion of 3-D location (r = -0.442). The results confirmed the correlated relationships among performance indexes and the importance of available satellite number and PDOP reflected location quality of 2-D and 3-D independently, but not reflected location error synthetically.

Keywords: global positioning system (GPS), location performance, satellite number, location error, location time, position dilution of precision (PDOP), GPS radio collar

Introduction

Global positioning system (GPS) radio telemetry has become an important wildlife research technique worldwide because of the obvious advantages of automated tracking of animal movements and the possible location accuracy (<10 m). The automated tracking realized the 24 hour continual location and reduces both the effort and cost for routine VHF (very high frequency) location. However, the raw data acquired through GPS radio collars contain bias and errors that must be addressed to arrive at accurate conclusions (Moen et al. 1996, Rempel and Rodgers 1997, D'Eon et al. 2002, Frair et al. 2004, D'Eon and Delparte 2005). Understanding, quantifying, and managing bias and errors in raw GPS radio-telemetry data sets requires much more work (D'Eon and Delparte 2005).

GPS was not designed for use under forest canopy (Edenius 1997). Therefore, before application in ecological research, the reliability of the technique must be determined by field test (Edenius 1997, Bowman et al. 2000, D'Eon and Delparte 2005). Satellite acquisition is the most important factor affecting fix rate and accuracy of ensuing locations (Moen et al. 1997). The number of satellites available to a GPS radio collar can be affected by physical obstructions between collars and satellites (Di Orio et al. 2003, Jiang et al. 2008). Researchers should identify potential effects that varying terrain, slope, and vegetation cover within their study area could have on GPS radio collar with respect to their study objectives synthetically (Biggs et al. 2001, Jiang et al. 2008).

Recent studies have shown that habitat features influence GPS location performance, such as satellite availability, fix rate, 2- or 3-dimensional (D) location proportion, postion dilution of precision (PDOP), and location error (Rempel and Rodgers 1997, Igota et al. 2002, Di Orio et al. 2003, Frair et al. 2004, Gau et al. 2004, Hensen and Riggs 2006, Jiang et al. 2008). Satellite availability and habitat features can affect both location accuracy and fix rate (Edenius 1997, Moen et al. 1997, Biggs et al. 2001, Hensen and Riggs 2006, Jiang et al. 2008). Rempel et al. (1995) found that fix rate was related positively to tree spacing and inversely to tree basal area and canopy closure in a coniferous plantation. Moen et al. (1996) and Rempel and Rodgers (1997) suggested that tree height is an important variable affecting fix rate, and Dussault et al. (1999) found a negative relationship between tree height and satellite number. Edenius (1997) and Hensen and Riggs (2006) concluded that fix rate was related inversely to canopy closure and basal area of stems in a boreal coniferous forest. Jiang et al. (2008) found that openness (the actual available sky percentage of a radio collar exposed to the sky when collar is blocked due to terrain and vegetation) and canopy closure affected location accuracy and most other performance indexs. Any combination of obstructions between collar and satellite decreased the available number of satellites, fix rate, increased the location time, and resulted in poorer location (higher PDOP, lower 3-D proportion, or greater location error) (Jiang et al. 2008). Additional information on location performance of GPS radio collar needs to be assessed under different habitat conditions and geographical areas (Moen et al. 1996, Biggs et al. 2001, D'Eon and Delparte 2005).

The GPS radio collar accumulated a series of index data related to the habitat and location accuracy, such as location time, PDOP, location data dimensions (2-D and 3-D), available and used satellite number. What are the relationships among these indexes, and, finally, how are they related to the location accuracy? To clarify these questions, we analyzed the differences of performance indexes recorded by the GPS radio collars among different location status (unsuccessful fix, successful 2-D, and 3-D location) and the relationships among indexes related to location quality.

Methods

Study Area

We carried out this study in northern Mount Fuji (35°22'–35°32'N, 138°33'–138°55'E), central Japan (Figure 1). Mount Fuji (3,776 m in elevation) is a volcano showing typical pyramid shape accompanying with wide surrounding foot area with gentle slope (Figure 1). Northern Mount Fuji is a special area with high variation in elevation from 800 to 3,776 m within about 18 km. The terrain changed from gentler to steeper with elevation increment and the mean slope degree ranges 0–31°. The vertical distribution of vegetation can be divided into montane, subalpine, and alpine zones.

Figure 1. The topography of Mount Fuji area and the distribution of test sites for testing the performance of GPS3300 radio collars (Lotek) during 2002–2004.

The plant communities in the montane and subalpine zones are mainly Corno-Fagetum crenatae, Abies homolepis, Rhodoreto-Pinetum densiflorae, Rhodoreto-Betuletum platyphyllae, and Abietum viechii-mariesii. In the area below 1,800 m, planted forest, golf ground, grassland and agricultural field are distributed in patches. More than 90% of forests in this area are dominated by coniferous species, with canopy closure varying from 0% in clearcut and open areas to about 80% in dense stands. Vegetation enters vegetative growth in early May and senesces at the mid-October.

Location Performance Tests

We employed 4 GPS3300 radio collars (Lotek Engineering Inc., Ontario, Canada) that consist of a 12-channel with differential correction. GPS antenna was embedded in belt and situated on the top of radio collar (Figure 2).

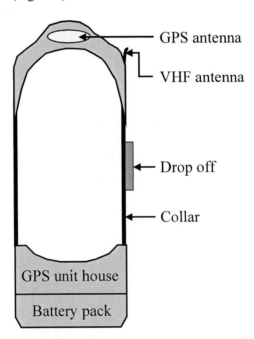

Figure 2. The illustration of GPS3300 radio collar (Lotek Engineering Inc., Ontario, Canada) used in the tests.

We tested the effects from terrain and vegetation on performance at 15 sites (Figure 1, Table 1) those represent the variations of elevation (800–2,500 m), aspect (northwest, north, and northeast), slope gradient (0–35°), and canopy closures (0–80%). We did the tests at all 15 sites in winter (February–March 2002–2004), and 3 forest sites of the 15 in summer (July–August 2003). For all tests, we hung collars vertically to a frame at 1 m aboveground and ensure GPS antenna in vertical position (Figure 3). For all tests we used 60-minute location interval, and left them for at least three days for ≥ 73 location attempts at each site.

Figure 3. The establishment of GPS3300 radio collars (Lotek) in northern Mount Fuji for testing collar location performance during 2002–2004.

At each testing site, we collected information related to nine habitat features (i.e. canopy closure, large tree basal area, shrub basal area, total tree basal area, large tree density, shrub density, total tree density, available sky, and openness) within a 10-m radius plot around a collar test site. We measured canopy closure in four cardinal directions and averaged the values. We measured the tree diameter of breast height (DBH) and recorded the numbers of large tree (DBH $\geq$ 10 cm) and shrub (DBH < 10 cm and height > 2 m), then calculated stem basal area (m^2/hectare), and tree density (number of tree per hectare). To examine the effects of the terrain, we calculated the available sky (the proportion of the sky theoretically available; obstructions were due to terrain but forest cover was disregarded) at all sites following the method of D'Eon et al. (2002). To examine the effects from the combination of terrain and vegetation, we created a new variable: "openness". We defined the openness as the percentage of available sky accessible to a GPS collar through direct line of sight in all directions and at all angles in a hemisphere without terrain and vegetation obstructions. At open sites, we measured the angles between the horizontal and the line from GPS collar to the mean top edge of the vegetation and terrain in the four cardinal directions. Then the openness was calculated using the mean surrounded angle (θ) as following equation: Openness (%) = (1 − sinθ)*100. In forest, we estimated percentage of available sky accessible to the collar in the four cardinal quarters of the hemispherical dome. Then we calculated mean percentage of available sky accessible to the collar (Jiang et al. 2008).

Table 1. Summary of habital features and location results of GPS3300 radio collars (Lotek) in northern Mount Fuji during the test period from 2002 to 2004

	Mount Fuji Aspect[a]	Elev. m	Slope °	n	Location time[c] s	Location error m	Success rate %	2-D[d] %	3-D[d] %	Satellite #	PDOP[f]	Openness %	Available sky %	Canopy cover %	Large tree density #/ha	Shrub density #/ha	Total tree density #/ha	Large tree basal area m²/ha	Shrub basal area m²/ha	Total tree basal area (m²/h)
	N	1020	0	73	49.4	10.9	100.0	1.4	98.6	7.0	3.1	88.4	92.5	0.0	0.0	0.0	0.0	0.0	0.0	0
	N	1020	2	122	63.2	12.0	97.9	6.3	91.7	5.8	4.2	63.8	92.5	0.0	0.0	0.0	0.0	0.0	0.0	0
Open flat	N	1430	7	122	50.5	9.3	100.0	0.0	100.0	7.0	3.2	85.7	85.8	0.0	0.0	0.0	0.0	0.0	0.0	0
	NE	1560	10	73	45.7	8.3	100.0	0.0	100.0	7.3	2.8	94.8	91.7	0.0	0.0	0.0	0.0	0.0	0.0	0
	NW	1510	4	73	49.7	8.0	100.0	0.0	100.0	7.0	3.2	90.9	89.6	0.0	0.0	0.0	0.0	0.0	0.0	0
	N	1520	21	73	65.9	19.5	98.6	19.2	79.5	5.5	5.3	46.3	79.9	0.0	0.0	0.0	0.0	0.0	0.0	0
Open slope	NW	1410	30, 19[b]	73	63.4	10.1	100.0	12.3	87.7	5.7	4.7	54.6	80.7	0.0	0.0	0.0	0.0	0.0	0.0	0
	N	2355	31	73	49.4	8.8	100.0	0.0	100.0	6.6	3.4	91.3	81.1	0.0	0.0	0.0	0.0	0.0	0.0	0
	N	1020	4	73	99.3	22.2	86.3	37.0	49.3	4.8	5.1	10.5	92.5	37.5	1177.7	5411.3	6589.0	61.3	10.6	71.9
	N	1020	4	74	117.1	37.9	52.7	33.8	18.9	3.9	4.6	6.5	92.5	66.3	1177.7	5411.3	6589.0	61.3	10.6	71.9
Closed flat	N	1500	10	92	92.8	55.8	87.0	60.9	26.1	4.1	6.1	6.8	80.8	66.3	891.3	668.5	1559.7	61.4	1.3	62.7
	NE	1560	11.5	73	81.1	28.6	91.4	47.9	42.5	4.4	4.9	8.5	89.2	47.5	668.5	1559.7	2228.2	53.2	3.1	56.3
	NW	1450	2	73	104.8	32.1	80.8	47.9	32.9	4.3	4.9	14.0	89.9	60.0	1145.9	1687.0	2833.0	64.2	3.3	67.5
	NW	1480	4.5	74	77.3	24.4	100.0	33.8	66.2	5.3	5.6	14.3	89.8	38.0	1273.2	63.7	1336.9	25.2	0.1	25.3
	N	2255	18	73	49.1	9.6	100.0	1.4	98.6	6.7	3.5	74.5	82.3	2.8	95.5	1527.9	1623.4	1.4	3.0	4.4
Closed slope	NW	1540	23	73	88.6	19.5	100.0	46.6	53.4	4.9	5.3	11.8	81.8	35.0	1018.6	3119.4	4138.0	43.3	6.1	49.4
	NW	1540	23	74	92.7	77.6	82.4	60.8	21.6	4.2	5.7	9.8	81.8	68.8	1018.6	3119.4	4138.0	43.3	6.1	49.4

[a] N, north; NE, northeast; NW, northwest. [b] at a joint between two slopes. [c] only included successful location. [d] D. dimension. [f] PDOP (position dilution of precision).

Data Analyses

We converted latitude/longitude of the location data to UTM zone 54 coordinates X, Y in the WGS84. For the location error, we adopted the method of Rempel et al. (1995), Moen et al. (1997), and Edenius (1997) for estimating 'truth location' and assumed that the mean of 3-D locations (n ≥ 10) at PDOP ≤ 5 to be true location for analytical purpose.

All statistic analyses were performed using SYSTAT 8.0 (SPSS 1998). We used Pearson correlation with associated Bonferroni probabilities to assess correlation between the performance indexes of GPS radio collars. We tested the differences among mean values of the performance index of different location status by analyses of variance (ANOVA), and pairwise mean differences by post Hoc test with Bonferroni adjustment. All statistical significance were set at the $P < 0.05$ level.

Results

Summary of Location

We made 1,361 attempts with 17 tests at 60-minute location interval at 15 sites with variable habitat features (Table 1, Figure 1). Fix rate ranged from 53–100%, and 3-D proportion ranged from 19–100% (Table 1). We found differences in all indexes (available satellite, location time, data dimension, and location error) among or between location status, but not in PDOP (Table 2). Available satellite number increased ($F_{2,\ 1358} = 715.267$, $P < 0.001$) from unsuccessful fix (3.0) to successful 2-D (4.1) and to 3-D (6.4) locations. Mean location time was 71.9 seconds, ranging from 27–160 (Table 2). In contrast with available satellite number, location time decreased ($F_{2,\ 1358} = 609.014$, $P < 0.001$) from unsuccessful fixes (160) to successful 2-D (96.7) and to 3-D (62.2) location. Mean location error was 23.2, ranging from 0.2–581.4 m (Table 1, 2, Figure 4).

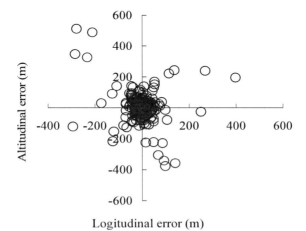

Figure 4. The distribution of location error (m) of GPS3300 radio collars (Lotek) in northern Mount Fuji during the test period from 2002 to 2004 (n = 1247).

The differences between location indexes reflected location error and showed that the error in 2-D location (47.8 m) was greater than in 3-D (13.6 m) ($F_{1, 1244}$ = 146.635, $P < 0.001$, Table 2). However, the location error difference between 2- and 3-D did not reflect PDOP, or there was no PDOP difference between 2-D and 3-D locations ($F_{1, 1244}$ = 715.267, $P = 0.143$, Table 2).

Table 2. The comparison of location performance indexes of GPS3300 radio collars (Lotek) among location status in northern Mount Fuji, Japan

			Mean	SD	Max	Min
Successful location	3-D (n = 899)	Satellite number	6.4 a	1.3	11.0	4.0
		Location time (second)	62.2 a	28.8	160.0	27.0
		PDOP [a]	4.3 a	3.0	24.2	1.5
		Location error (m)	13.6 a	22.3	354.5	0.2
	2-D (n = 348)	Satellite number	4.1 b	0.8	6.0	3.0
		Location time (second)	96.7 b	37.6	160.0	27.0
		PDOP [a]	4.6 a	3.5	20.6	1.6
		Location error (m)	47.8 b	76.8	581.4	0.5
Unsuccessful location	No location (n = 114)	Satellite number	3.0 c	1.0	5.0	0.0
		Location time (second)	160 c	0	160	160

Values within same category in mean column followed by the same letter do not differ (P>0.05). [a] PDOP (position dilution of precision).

Relationships among Location Performance Indexes

We found that the available satellite number affected location time (r = -0.390, $p < 0.001$), location data dimension (r = 0.639, $p < 0.001$), PDOP (r = -0.362, $p < 0.001$), and finally location error (r = -0.282, $p < 0.001$) (Table 3, Figure 5, A – D). The location time, PDOP, and location error decreased with the increasing available satellite number. More satellite number resulted in more 3-D location. Meanwhile, small PDOP reflect less location error (r = 0.407, $p < 0.001$), less location time (r = 0.088, $p < 0.018$, Table 3, Figure 5, E – F). However, PDOP did not show relation with the proportion of location data dimension (r = 0.042, $p = 1.000$, Table 3, Figure 5, G). Further, longer location time correlated with greater location error (r = 0.196, $p < 0.001$) and less proportion of 3-D location (r = -0.442, $p < 0.001$, Table 3, Figure 5, 3, H – I). Finally, 3-D location reflect less location error than 2-D (r = -0.325, $p < 0.001$) (Table 3, Figure 5, J).

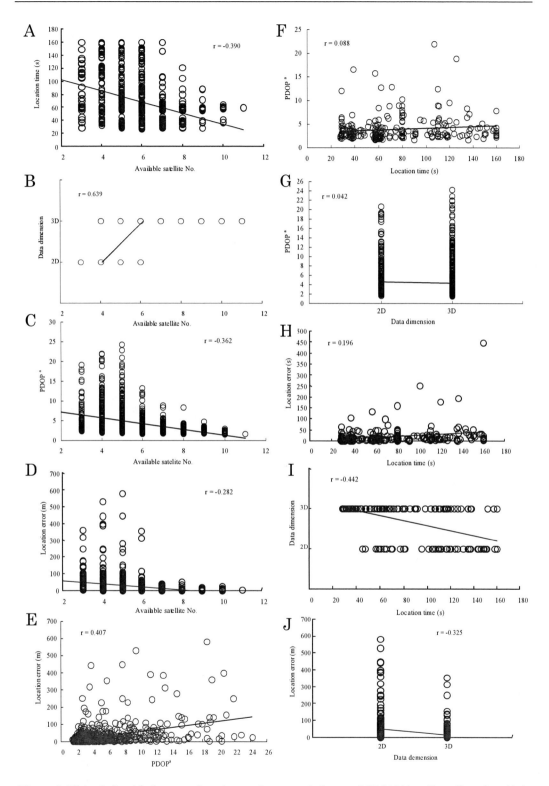

Figure 5. The relationship between location performance indexes of GPS3300 radio collars (Lotek) in northern Mount Fuji, central Japan (n = 1247). [a] PDOP (position dilution of precision).

Table 3. The correlation between location performance indexs of GPS3300 radio collars (Lotek) in northern Mount Fuji, central Japan (n=1247)

Matrix of Pearson correlation

	Available satellite #	Location time	Data dimension	Location error	PDOP [a]
Available satellite #	1				
Location time	-0.390	1			
Data dimension	0.639	-0.442	1		
Location error	-0.282	0.196	-0.325	1	
PDOP [a]	-0.362	0.088	-0.042	0.407	1

Matrix of Bonferroni probabilities

	Available satellite #	Location time	Data dimension	Location error	PDOP [a]
Available satellite #	0.000				
Location time	0.000	0.000			
Data dimension	0.000	0.000	0.000		
Location error	0.000	0.000	0.000	0.000	
PDOP [a]	0.000	0.018	1.000	0.000	0.000

[a] PDOP (position dilution of precision).

Discussion

A minimum of 3 satellites are required to obtain a 2-D location, and 4 are required to obtain a 3-D location, which is more accurate than a 2-D location, within a short period of time (Moen et al. 1996, Rodgers et al. 1996, D'Eon et al. 2002). Available satellite number increasing from unsuccessful fix, to successful 2-D, and to 3-D locations confirmed the importance of available satellite number for a successful and high quality location. This importance was confirmed by the positive relationship between satellite number and proportion of 3-D location, and the negative relationship between location error, PDOP and available satellite number.

Fix rate ranging from 53–100% and 3-D proportion ranging from 19–100% reflect the negative relationship between fix rate and the influence from vegetation and terrain obstructions (Edenius 1997, D'Eon et al. 2002, Di Orio et al. 2003, Hansen and Riggs 2006, D'Eon and Delparte 2005, Hansen and Riggs 2006, Jiang et al. 2008). A negative relationship between openness and PDOP confirmed the negative effects on location quality from habitat features (Jiang et al. 2008) because the collar could not reach enough satellites, or a good enough satellite signal to calculate a location under the disturbance of vegetation and topographical obstructions (Rodgers et al. 1996, Edenius 1997, Moen et al. 1997, Dussault et al. 1999, Hansen and Riggs 2006).

Location time for each attempt decreased from unsuccessful fix to successful 2-D and to 3-D locations. Any disturbance between collars and satellites in habitat made the location time longer, or there were not enough available satellites for calculating a 3-D location, resulting in poor location with greater PDOP, or unsuccessful fix (Jiang et al. 2008). This was confirmed by the negative correlation between location time and available satellite number, and the relationship suggests that the available satellite number not only decided the location quality, but also influenced the longevity of the collar battery. This is important for field researchers in designing their location schedule by considering the study purpose, habitat condition, and the longevity of collar battery synthetically. The longer location time due to increasing canopy closure has been reported by Hansen and Riggs (2006) and Jiang et al. (2008).

The positive relationship between PDOP and location time was reasonable because low PDOP means high quality location, and high quality location resulted from a higher available satellite number, and greater available satellite number related to less location time. This was also confirmed by the positive relation between location time and location error and the negative relation between location time and the proportion of 3-D location.

The differences between location indexes reflected location error and showed that the error in 2-D location was greater than in 3-D. This was confirmed by the negative relationship between the location error and the proportion of 3-D location. In general, 3-D location reflected less location error than 2-D (Rempel and Rodgers 1997, Edenius 1997, D'Eon and Delparte 2005, Jinag et al. 2008).

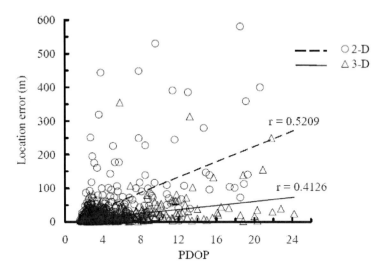

Figure 6. The relationship between PDOP (position dilution of precision) and 2-D (dimension), 3-D location error, respectively, showing the independence of the two relationships between 2-D and 3-D (n = 348 for 2-D, n = 899 for 3-D) of GPS3300 radio collars (Lotek) in northern Mount Fuji, central Japan.

The negative relations of PDOP, location error, and the proportion of 2-D location with available satellite number suggest that the available satellite number decides the location quality. Edenius (1997), Hansen and Riggs (2008), and Jiang et al. (2008) pointed out that the accuracy varied inversely with canopy closure and basal area of stems. In general, 3-D locations with smaller PDOP are better than 2-D locations with greater PDOP (D'Eon and Delparte 2005, Jiang et al. 2008). This is expected because PDOP is a direct location quality index, and small PDOP means high-quality location.

D'Eon and Delparte (2005) confirmed that smaller PDOP reflects less location error. Rodgers et al. (1996), D'Eon et al. (2002), and Jiang et al. (2008) confirmed that 3-D location reflects less location error. However, PDOP did not relate to the proportion of location data dimension, or no PDOP difference between 2-D and 3-D locations. This is unexpected because both PDOP and location dimension are the direct indexs of location quality. This unexpected relationship suggests that PDOP may reflect location quality of 2-D and 3-D independently, but not reflect location error synthetically (Figure 6). Also, PDOP may reflect the dilution of precision of satellite signal mainly, but not the final result of location accuracy of 2-D and 3-D differentially.

Acknowledgements

We are grateful to A. Fujisono for her help in the field, and to H. Ueda for his constructive comments and suggestions regarding the research design and statistical analyses. We also appreciate the Japan Society for the Promotion of Science for funds for this research. This research was partially supported by the Ministry of Education, Culture, Sports, Science and Technology, Japan, Grant-in-Aid for Scientific Research, Basic Research (B), No. 14360198, 2002.

References

Biggs JR, Bennett KD, Fresquez R (2001) Relationship between home range characteristics and the probability of obtaining successful global positioning system (GPS) collar positions for elk in New Mexico. *Western North American Naturalist* **61**:213–222

Bowman JL, Kochanny CO, Demarais S, Leopold BD (2000) Evaluation of GPS collar for white-tailed deer. *Wildl. Soc. Bull.* **28**:141–145

D'Eon RG, Serrouya R, Smith G, Kochanny CO (2002) GPS radiotelemetry error and bias in mountainous terrain. *Wildl. Soc. Bull.* **30**:430–439

D'Eon RG, Delparte D (2005) Effect of radio-collar position and orientation on GPS radio-collar performance, and the implications of PDOP in data screening. *J. Appl. Ecol.* **42**:383–388

Di Orio AP, Callas R, Schaefer RJ (2003) Performance of two GPS telemetry collars under different habitat conditions. *Wildl. Soc. Bull.* **31**:372–379

Dussault C, Courtois R, Ouellet JP, Huot J (1999) Evaluation of GPS telemetry collar performance for habitat studies in the boreal forest. *Wildl. Soc. Bull.* **27**:965–972

Edenius L (1997) Field test of a GPS location system for moose *Alces alces* under Scandinavian boreal conditions. *Wildl. Biol.* **3**:39–43

Frair JL, Nielsen SE, Merrill EH, Lele SR, Boyce MS, Munro RHM, Stenhouse GB, Beyer HL (2004) Removing GPS collar bias in habitat selection studies. *J. Appl. Ecol.* **41**:201–212

Gau RJ, Mulders R, Ciarniello LM, Heard DC, Chetkiewicz C-LB, Boyce M, Munro R, Stenhouse G, Chruszcz B, Gibeau ML, Milakovic B, Parker KL (2004) Uncontrolled field performance of Televilt GPS-SimplexTM collars on grizzly bear in western and northern Canada. *Wildl. Soc. Bull.* **32**:693–701

Hansen MC, Riggs RA (2006) Accuracy, precision, and observation rates of global positioning system telemetry collars. *J. Wildl. Manage* **72**:518-526.

Igota H, Waseda K, Sakuragi M, Uno H, Kaji K, Kaneko M, Akamatsu R, Maekawa K (2002) Evaluation of GPS telemetry and its application for Hokkaido sika deer. *Sci. Mammal.* 42:113–121 (in Japanese with English abstract)

Jiang ZW, Sugita M, Kitahara M, Takatsuki S, Goto G, Yoshida Y (2008) Effects of habitat feature, antenna position, movement, and fix interval on GPS radio-collar performance in Mount Fuji, central Japan. *Ecol. Res.* **23**(3): 581-588

Moen R, Pastor J, Cohen Y (1997) Accuracy of GPS telemetry collar locations with differential correction. *J. Wildl. Manage* **61**:530–539

Moen R, Pastor J, Cohen Y, Schwartz CC (1996) Effects of moose movement and habitat use on GPS collar performance. *J. Wildl. Manage* **60**:659–668

Rempel RS, Rodgers AR (1997) Effects of differential correction on accuracy of a GPS animal location system. *J. Wildl. Manage* **61**:525–530

Rempel RS, Rodgers AR, Abraham KF (1995) Performance of a GPS animal location system under boreal forest canopy. *J. Wildl. Manage* **59**:543–551

Rodgers AR, Rempel RS, Abraham KF (1996) A GPS-based telemetry system. *Wildl. Soc. Bull.* **24**:559–566

SPSS (1998) SYSTAT 8.0 for windows. SPSS, Chicago, Illinois, USA.

In: Global Positioning Systems
Editors: V. Asphaug and E. Sørensen, pp. 57-60

ISBN: 978-1-60741-012-6
© 2010 Nova Science Publishers, Inc.

Chapter 4

"APPLICATION OF A GEOGRAPHICAL INFORMATION SYSTEM (GIS) AND THE GLOBAL POSITIONING SYSTEM (GPS) TO DENGUE VIRUS VECTOR: *AEDES* MOSQUITOES DISTRIBUTION IN AN EPIDEMIC AREA OF THAILAND", A TECHNICAL COMMENT

Viroj Wiwanitkit
Wiwanitkit House, Bangkhae, Bangkok, Thailand

Abstract

In Thailand, dengue hemorrhagic fever is an important infectious disease. There are several outbreaks of this infection in each decade. This fact causes dengue hemorrhagic fever to be a major public health problem. As a rule, vector survey is a useful basic preventive technique for surveillance and control of mosquitoes. The author has ever published the previous report on dengue hemorrhagic fever using a global positioning technique. The objective of that quoted study is to apply the Geographical Information System (GIS) and the Global Positioning System (GPS) to the dengue virus vector, *Aedes* mosquitoes, distribution in an epidemic area of Thailand. Here, the author provides a technical comment based on that report.

Keywords: GIS, GPS, dengue, mosquito, technical comment

Introduction

In Thailand, dengue hemorrhagic fever is an important infectious disease. There are several outbreaks of this infection in each decade [1]. This fact causes dengue hemorrhagic fever to be a major public health problem [1]. There were thousands of cases of DF/DHF documented by the Epidemiological Division, Ministry of Public Health during the past decade [2]. In this decade, the number of reported infected cases from the central region of Thailand was the highest, followed by the northeastern region [3].

Focusing on mode of transmission, dengue virus, are routinely transmitted to humans through vector transmission mode. The bite of infective *Aedes* mosquitoes, namely *Aedes aegypti* , that are common in urban areas, during intermittent rainfall in these tropical regions [4]. In Thailand, *Aedes aegypti* is the primary most common vector involved in dengue virus outbreaks following by *Aedes albopictus* [5].

As a rule, vector survey is a useful basic preventive technique for surveillance and control of mosquitoes. The author has ever published the previous report on dengue hemorrhagic fever using a global positioning technique. The objective of that quoted study is to apply the Geographical Information System (GIS) and the Global Positioning System (GPS) to the dengue virus vector, *Aedes* mosquitoes, distribution in an epidemic area of Thailand [6]. Here, the author provides a technical comment based on that report.

Technical Comment

1. Study Design

The study can be classified as a kind of research on global positioning systems. Usually, this kind of study makes of both wet laboratory and dry laboratory. Focusing on the wet laboratory, the field experimental research is the core. In the quoted study [6], the focused area is the Chombung District, Ratchaburi province areas in the western Region of Thailand. The things to be included in the field experimental research include the field collection of mosquitoes following by the Identification of Mosquitoes.

For the dry laboratory, the informatics research is usually used. The geographical information system (GIS) is the effective tool that is widely used in the present day. The Global Positioning System (GPS) instruments, such as TDSI, are applicable at this point. Basically, differentially processed GPS data will be further incorporated into the GIS pattern for analysis and mapping. In general, the important data include the study areas, including village name, house number, and number of adult mosquitoes collected. Both rover and base station units will be further processed simultaneously to allow differential correction of rover data using specific GPS software.

2. Results

Important results from this kind of study include GIS data and map. These results are usually shown in the pattern of map. In Figure 1, the author shows the direct copied of figure in the previous report to be an example for the reader on this kind of map. In the figure 1, the GPS data were subjected to differential processing to create a map with a spatial resolution of one meter then imported into a GIS and a new database created that included village name and house number, and mosquito population.

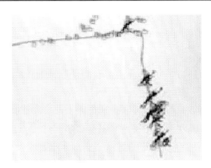

Figure 1. Modified figure from the previous publication, "Portion of GIS map and database for Moo 3 and Moo 6 Tambon Chombung, Chombung District, Ratchaburi Province, Thailand. Distribution of *Aedes* mosquitoes plotted layer in a GIS. [6]" The red dot represents the cluster of houses and the black mosquito represents the volume of amount of mosquito vectors.

3. Discussion

An important prevention and control of DF/DHF outbreaks is the control of the vector mosquitoes Ae. aegypti and Ae. albopictus. GPS, a device used to measure the precise position on the earth in order to detect crustal movement, can be useful for this purpose. Basically, GPS data, ,an organized collection of computer hardware, software, geographic data, and personnel to efficiently capture, store, update, manipulate, analyze, and display all forms of geographically referenced information, can also additional used for meteorological study. These two techniques can be helpful for vector surveillance and further infection control.

These systems can be effectively used in the surveillance and monitoring of vector-borne diseases including dengue [7–10]. These tools can be useful for predicting changes in habitats of mosquito vectors as they affect disease transmission [11-14]. The combination between GIS and GPS improves the quality of spatial and nonspatial data for analysis and decision making by providing an integrated approach to dengue control and surveillance at the local, regional and/or national level. According to the quoted study [6], a GIS and a new database including included village name, house number, demographic data on house occupants, adult prone virus-infected *Aedes aegypti* populations and seroepidemiology data on house occupants were finalized derived and could be further used as powerful tools to monitor the status of efforts to control *Ae. aegypti* breeding sites and to evaluate the impact of this control effect on dengue and DHF transmission.

References

[1] *Annual epidemiological surveillance report*. Ministry of Public Health Organization, Thailand. 1996
[2] *Annual epidemiological surveillance report*. Ministry of Public Health Organization, Thailand. 1997.
[3] *Annual epidemiological surveillance report*. Ministry of Public Health Organization, Thailand. 1999.

[4] Fort Collins, Co. Duane J. Gubler DJ., Sc. D and Edward B. Hayes, MD. Dengue and dengue hemorrhagic fever. Dengue branch and the division of vector-borne infectious diseases, *Centers for disease control, CID.* 1992.

[5] *Annual epidemiological surveillance report.* Ministry of Public Health Organization, Thailand. 2000.

[6] Srisupanant M, Sithiprasasna R, Patpoparn S, Attatippaholkun W, Wiwanitkit V. Application of a geographical information system (GIS) and the global positioning system (GPS) to dengue virus vector: *Aedes* mosquitoes distribution in an epidemic area of Thailand. In: Wiwanitkit, ed. Focus on Climate Change and Health. New York: Nova: 2009

[7] Glass GE, Schwartz BS, Morgan JM.III, Johnson DT, Noy PM, Israel E. Environmental risk factors for Lyme disease identified with geographic information system. *Am J Public Health 1995*;85:944-948.

[8] Beck LR, Rodrigues MH, Dister SW, Rodrigues AD, Rejmankova E, Ulloa A, et al. Remote sensing as a landscape epidemiologic tool to identify village at high risk for malaria transmission. *Am J Trop Med Hyg* 1994;51:271-280.

[9] Richards FO, Jr. Use of geographic information systems in control programs for onchocerciasis in Guatamala. *Bull Pan Am Health Organ* 1993;27:52-55.

[10] Nualchawee K, Singhasivanon P, Thimasam K Doereang D, Linthicum K, Sithiprasasna R and Rajbhandari PL. Correlation between malaria incidence and changes in vegetation cover using satellite remote sensing and GIS techniques. In: *Proceedings of the International Geoscience and Remote Sensing Symposium (IGARSS), August 1997.* Institute of Electrical and Electronic Engineers, New York (in press).

[11] Hayes RO, Maxwell EL, Mitchell CJ and Woodzick TL. Detection, identification and classification of mosquito larval habitats using remote sensing scanners in earth-orbiting satellites. *Bull WHO.1985*;63:361-374.

[12] Linthicum KJ, Bailey CL, Davies FG and Tucker CJ. Detection of Rift valley fever viral activity in Kenya by satellite remote sensing imagery. *Science 1987*;235:1656-1659.

[13] Linthicum KJ, Bailey CL, Tucker CJ. Gordon SW, Logan TM, Peters CJ and Digoutte JP. Observations with NOAA and SPOT satellites on the effect of man-made alterations in ecology of the Senegal River basin in Maurtania on Rift Valley fever virus transmission. *Sistema Terra* 1994;3:44-47.

[14] Pope KO, Sheffner EJ, Linthicum KJ, Bailey CL, Logan TM, Kasischke ES, Birney K, Njou AR and Roberts CR. Identification of central Kenyan Rift Valley fever virus vector habitats with Landsat TM and evaluation of their flooding status with airbone image radar. *Remote Sensing of the Environment 1992*;40:185-196.

In: Global Positioning Systems
Editors: V. Asphaug and E. Sørensen, pp. 61-116

ISBN 978-1-60741-012-6
© 2010 Nova Science Publishers, Inc.

Chapter 5

ACCURATE GPS-BASED GUIDANCE OF AGRICULTURAL VEHICLES OPERATING ON SLIPPERY GROUND

Benoit Thuilot[1],*, *Roland Lenain*[2],†
Philippe Martinet[1] *and Christophe Cariou*[2]
[1]LASMEA, Clermont-Ferrand University, 24 avenue des Landais,
63177 Aubière Cedex FRANCE
[2]CEMAGREF Institute, 24 avenue des Landais - BP 50085
63172 Aubière Cedex FRANCE

Abstract

The development of automatic guidance systems for agricultural vehicles is receiving considerable attention from both researchers and manufacturers. The motivations in such automated devices are, on one hand, to reduce considerably the arduous driving task, and on the other hand, to improve the efficiency and the quality of the agronomic work carried out.

Such guidance devices require realtime vehicle localization on an unstructured area, such as agricultural fields. Nowadays, RTK GPS sensor appears as a very suitable sensor for these applications, since it can supply this information with a satisfactory centimeter accuracy at a high 10 *Hz* frequency, without requiring any preliminary equipment of the field.

In this chapter, it is demonstrated that very accurate curved path following can actually be achieved by agricultural vehicles, even on slippery ground, relying on a single RTK GPS sensor.

In a first step, in order to benefit from recent advances in Control Theory, sliding effects have been omitted, and therefore guidance laws have first been designed relying on a vehicle kinematic model. More precisely, taking advantage from structural properties of these models (they can be converted into a so-called chained form), curved path following has been achieved by designing a non-linear control law. Full-scale experiments reveal a very satisfactory guidance accuracy, except when the vehicle enters into sharp curves or when it moves on sloping fields.

---

*E-mail address: Benoit.Thuilot@lasmea.univ-bpclermont.fr
†E-mail address: Roland.Lenain@cemagref.fr

In these two latter situations, guidance accuracy is damaged since the vehicle undergoes sliding effects. In agricultural applications, vehicle dynamic models appear untractable from the control design point of view. Therefore, it is here proposed to describe sliding effects as a structured perturbation acting on vehicle kinematic model. Adaptive Control framework, and more precisely Internal Model techniques, can then be used jointly with the above mentioned non-linear control law to reject sliding effects, when still preserving all the advantages of the previously designed non-linear guidance law.

Experiments demonstrate satisfactory guidance accuracy when the vehicle moves along a slope or when it executes sharp curves, excepted at their beginning or end. These transient guidance errors mainly ensue from delays introduced by the actuation device. Since the shape of the path to be followed is known, beginning/end of curves or slopes can be anticipated. Model Predictive Control framework is here used to provide such an anticipation. Satisfactory experimental results display the performances of the overall control scheme.

1. Introduction

The development of guidance systems for agricultural vehicles receives more and more attention from researchers and manufacturers. The objectives and the motivations are numerous, since automatic guidance:

- reduces the work arduousness: for instance, achieving perfectly parallel runs when driving manually, is very tiring over hours.

- allows the driver to fully devote his time to the monitoring and the tuning of the tool. This clearly can improve the quality of the agronomic work carried out.

- ensures an optimal work precision all day long and on the whole field. Double applied and skipped areas between successive passages can then be minimized, so that the exact placement of field inputs (seeds, fertilizers, pesticides, etc.) can be achieved. Their cost can then be reduced, and their environmental-unfriendly features (e.g. when pesticides are considered) can be supervised.

- allows to operate wider tools at higher speeds. Therefore productivity can be increased.

Such guidance devices require realtime vehicle localization on an unstructured area, such as agricultural fields. Nowadays, it can be supplied very satisfactorily by an *RTK-GPS sensor (RealTime Kinematic - Global Positioning System)*: absolute 3D-localization information is available at a 10 *Hz* frequency with a centimeter accuracy. This frequency, as well as this accuracy, are clearly suitable for vehicle guidance applications.

This sensor appears very attractive, since no preliminary field equipment is required (there is no need for buried-cables, magnets or beacons sets, etc., such as with some other localization devices), and information can be provided whatever the weather or the light conditions (localization can be obtained by night, in presence of dust, etc., which is not the case for instance with a camera sensor). Moreover, this sensor is here fully reliable, since interruptions in GPS signal reception, which is one major concern, do not occur in agricultural tasks where vehicles move on open fields.

Therefore, the RTK GPS sensor is now frequently considered to address vehicles guidance applications. It is usually the keystone of a perception device enclosing multiple sensors. For instance, in agricultural literature, guidance applications have been reported relying on a GPS and a fiber optic gyroscope (FOG) [12], a GPS, a camera and near infrared reflectance sensors [16], or multiple GPS antennas [13]. Similar approaches are also developed in Japanese universities [20].

Moreover, some companies are already marketing some guidance systems relying on an RTK GPS sensor. The first commercial device dedicated to agricultural use, has been introduced in 1997 by the Australian company *AgSystems*. This device, named *BEELINE Navigator*, consists in an RTK GPS sensor coupled with an Inertial Navigation System (INS), and is mainly dedicated to achieve perfect straight runs. In the U.S.A., this market is currently led by the company *Novariant* (previously *IntegriNautics*). Their *AutoSteer System* relies on a 3-antennas RTK GPS sensor, that can provide the whole tractor attitude. Therefore, curved path following can also be addressed. GPS systems suppliers as well as agricultural manufacturers are also investing in this market: *Trimble* is selling *the AgGPS Autopilot*, relying mainly on their *AgGPS 214* and now *AgGPS 252* RTK-GPS sensors, when manufacturer *John Deere* has established partnerships with the university of Illinois and Stanford university, and is now marketing *the AutoTrac* system, relying on their *StarFire* RTK GPS receiver. These two systems are also mainly dedicated to achieve straight line following. Additional inertia sensors allow to compensate GPS information for slope.

These research works and commercial products show clearly the interests in vehicle automatic guidance dedicated to agricultural tasks, and the relevancy of the RTK GPS sensor to address such applications.

Automatic guidance of agricultural vehicles is then addressed here, with the aim to propose enhanced capabilities, and relying nevertheless solely on an 1-antenna RTK GPS sensor.

Firstly, it has been above mentioned that most of marketed devices are currently devoted to applications where the vehicles must execute perfectly straight lines (e.g. row cropping, harvesting, . . .). Extending guidance systems capabilities in order that the vehicles could also follow curved paths would be of practical interest (in order to achieve automatic half-turns, field boundaries following, etc.).

Secondly, if the vehicles are expected to achieve curved paths on agricultural fields, they will inevitably undergo sliding effects. Therefore, if high accurate guidance is the objective, it appears necessary to characterize the influence of such effects on the vehicle motion, and to account for sliding phenomenon in guidance laws. The same conclusion has also to be derived if the vehicles are expected to perform straight lines on sloping fields.

Finally, it can also be observed that most of the above-mentioned guidance systems either described in the literature, or already marketed, rely on several sensors. Such an equipment is efficient since it provides control designers with numerous information on the vehicle attitude (or even with the whole vehicle attitude if a 3-antennas RTK GPS sensor is considered), but is quite expensive. It is therefore interesting to investigate first the guidance accuracy that could be expected when relying solely on an 1-antenna RTK GPS sensor.

The application considered here is then vehicles guidance along arbitrary curved paths, on agricultural fields where sliding is likely to occur. Our aim is to achieve high accuracy guidance relying solely on a 1-antenna RTK GPS sensor.

The approach presented here relies, on one hand on modeling developments, and on the other hand on control design refinements.

Since sliding effects on the vehicle motion have to be accounted, it would seem natural to rely on vehicle dynamic models. However, the description of dynamic features leads to very large models. Moreover, these models encompass numerous parameters (masses, spring stiffnesses, etc.) whose values are badly known, and very difficult to reach through experimental identification. Finally, some parameters that are of crucial importance (e.g. wheels-ground contact conditions) are even continually varying. Therefore, addressing vehicles control from dynamic models proposes serious difficulties, both from the theoretical and the implementation points of view. Few works (e.g. [5]-[19]) have been reported, but they are concerned with conventional car control on asphalted ground, and moreover, their aim is to prevent the vehicles from sliding. Agricultural context is more involved: the vehicles necessarily undergo sliding, since they are moving on irregular slippery ground. In such harsh conditions, vehicle dynamic models do not appear very tractable from the control design point of view. Therefore, it is proposed here to still rely on standard vehicle kinematic models, derived under pure rolling and non-sliding conditions at wheels-ground contact points, and to account for sliding effects by introducing an additive structured perturbation. Experiments reported here display that these much more compact models, named here below *extended kinematic models*, can nevertheless accurately account for sliding effects on vehicle motion.

This vehicle modeling provides with the very attractive feature that control objective can be addressed, by still relying on powerful guidance laws that can be designed when it is assumed that the vehicles do not slide, while sliding effects can be accounted by introducing adaptive control techniques. More precisely, it is shown here that accurate curved path following can be achieved, by relying on non-linear control laws inferred, under non-sliding assumption, from *Chained Systems Theory* (e.g. [17]), and then extended according to *Internal Model Adaptive* techniques (e.g. [2]-[11]) in order to account for sliding effects. The only unsatisfactory situations are when sliding conditions present large variations (e.g. at the beginning/end of a curve): transient guidance errors can then be observed, due to the delays enclosed in the guidance feedback loop (originating mostly from actuator features). It is however shown here that, by introducing *Model Predictive Control* techniques (e.g. [14]-[15]) into vehicles guidance laws, guidance accuracy can also be preserved in such situations. Relevancy of this overall control scheme is supported by numerous satisfactory experimental results.

This chapter is organized as follows: first, the experimental context is introduced in Section 2.. Then, the proposed vehicle modeling is described in Section 3.. Next, guidance laws design is detailed in Section 4.. Finally, capabilities of both vehicle modeling and vehicle guidance laws are discussed, relying on numerous experimental reports, in Section 5..

2. Experimental Context

All the vehicle guidance laws designed in the forthcoming Section 4. have been implemented on a full-scale vehicle, and their capabilities have always been extensively investigated via numerous experiments. The more significant experimental results are reported in

Section 5.

More precisely, experiments have been performed on *a commercial ARES 640 Renault-Agriculture* farm tractor, lent by *CLAAS*, since this research work has been carried out in partnership with this German agricultural vehicles manufacturer.

Figure 1. ARES 640 farm tractor.

The main technical features of this tractor, shown on Figure 1, are given in Table 1. No technical modification has been achieved on this commercial farm tractor, excepted the addition of a *Danfoss* electro-hydraulic valve that enables the automatic steering of the front wheels. The features of this actuator are more detailed in forthcoming Section 4.3., since it will be shown that they have to be accounted in the control law design in order that guidance tasks could be achieved with a satisfactory accuracy. It can also be mentioned that in the addressed guidance applications, the tractor velocity is never automatically controlled. Therefore no specific actuator has been introduced: in all experiments, the tractor velocity is adjusted manually by the farmer.

Table 1. ARES 640 main technical features

Dimensions:		*Performances:*	
Length:	5.2 m	Maximum velocity:	37.5 $km.h^{-1}$
Wheelbase:	2.75 m	Working velocity:	8-10 $km.h^{-1}$
Height:	2.95 m	Horsepower:	130 CV
Weight:		Maximum Torque:	54.5 daN.m
without implement:	6370 Kg		

The tractor localization is achieved relying on an RTK GPS device manufactured by *Thales-Navigation*. More precisely, it is a dual frequency *"Aquarius5002"* unit. When position measurements are delivered with the upper sampling frequency $f_s = 10\ Hz$, as it will be in all forthcoming experiments, the claimed accuracy is 2 centimeters. A 1 centimeter accuracy could even be obtained with this unit, but only with a 1 *Hz* sampling frequency, which is inconsistent with vehicle guidance applications.

RTK GPS equipment is shown on Figures 2. The tractor GPS antenna, shown on Fig-

(a) Tractor GPS antenna   (b) Base GPS station

Figure 2. RTK GPS equipment.

ure 2(a) is located on the top of the tractor cabin, straight up above the center of the rear axle. This last requirement, as it will be detailed in forthcoming Section 3.4.1., is crucial in order to provide vehicle guidance laws with exactly the position measurements that are expected. Moreover, this position is also consistent with the technological requirements: the top of the tractor cabin is the highest part of the vehicle, and therefore the most convenient place to see as many satellites as possible. The GPS sensor on-boarded on the tractor receives also, via UHF link, the correction information provided online by the GPS base station, depicted on Figure 2(b). This GPS antenna is located accurately at a point whose absolute position is perfectly known, in order that corrections as accurate as possible can be sent. The quality of these corrections is crucial to enable the mobile GPS receiver to deliver the tractor position measurements with the expected 2 cm accuracy.

The on-boarded control device consists in a Pentium based computer. The vehicle position measurements, supplied by the GPS unit, are received via a first RS232 link. They are proceeded by the vehicle guidance laws, implemented in C++ language. The computed steering angle value is then sent to the steering actuator via a second RS232 link. The farmer can communicate with this control device, or simply supervise the accuracy of the running guidance task, via a Graphical User Interface, shown on Figure 3.

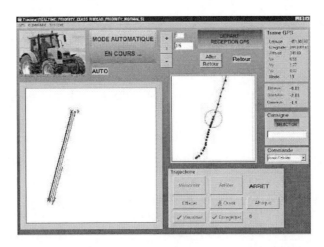

Figure 3. On-boarded Graphical User Interface.

3. Vehicle Modeling

The aim of this section is to derive a farm tractor model from which control laws achieving accurate guidance could conveniently be designed. In system modeling, a compromise is generally faced between:

- on one hand, a model which finely describes system behavior, but whose complexity forbids control law design,

- on the other hand, a very simple model, easy to manage from a control design point of view, but which imperfectly accounts for system behavior.

When considering agricultural vehicles, dynamic models appear untractable from a control design point of view. Therefore, as it has been mentioned in the *Introduction*, tractor guidance laws are here designed from a kinematic model, that has been extended in order to account for the main dynamic effects in agricultural applications, namely sliding effects.

More precisely, this section is organized as follows: modeling assumptions and notations are first presented in Section 3.1.. Next, the derivation of a vehicle kinematic model under standard non-sliding assumptions is recalled in Section 3.2.. Extension of this kinematic model, in order to account for sliding, is then presented in Section 3.3.. Finally, measurement and/or estimation of the vehicle state variables is discussed in Section 3.4..

3.1. Modeling Assumptions and Notations

The farm tractor and its possible implement are assumed to constitute a single rigid body (this is the case e.g. with seeders, sprayers, ...). They are here described according to a *bicycle model* (also named *Ackermann's model*): front and rear axles are both replaced by single virtual wheels, located at mid-distance between the actual wheels, see Figure 4. This assumption is quite common in mobile robots literature as long as the vehicles do not slide, see for instance [21]. Its relevancy, when control design is addressed, has been established a posteriori by numerous satisfactory experimental reports. Experiments reported in forthcoming Section 5. show that it is still relevant in presence of sliding, when sliding effects are accounted as a structured perturbation into the vehicle kinematic model.

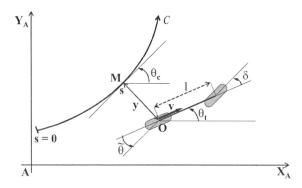

Figure 4. Vehicle description.

Since the control objective is path following, the vehicle configuration is here described with respect to the reference path C, rather than with respect to an absolute reference frame. More precisely, the notations are, see also Figure 4:

- C is the path to be followed. It is defined in an absolute frame $[A, X_A, Y_A)$,

- O is the center of the vehicle virtual rear wheel (i.e. the center of the tractor rear axle),

- M is the point on C which is the closest to O. M is assumed to be unique. In practical situations, this assumption is satisfied since, on one hand the tractor remains always close to C, and on the other hand the curvature of path C is small.

- s is the curvilinear abscissa of point M along C, $c(s)$ denotes the curvature of path C at that point, and $\theta_c(s)$ stands for the orientation of the tangent to C at that point, with respect to frame $[A, X_A, Y_A)$,

- θ_t is the orientation of the tractor centerline with respect to frame $[A, X_A, Y_A)$. Therefore $\tilde{\theta} = \theta_t - \theta_c(s)$ denotes the angular deviation of the tractor with respect to path C.

- y is the lateral deviation of the tractor with respect to C,

- v is the tractor linear velocity at point O,

- δ is the orientation of the front wheel with respect to tractor centerline,

- l is the tractor wheelbase.

Since the farm tractor and its possible implement are considered as a single rigid body, the vehicle configuration is described without ambiguity when the location of any point of the vehicle, for instance O, and vehicle centerline orientation are both given. Relying on above notations, these two information can respectively be represented by the couple (s, y) and the variable $\tilde{\theta}$. Vehicle *state vector* can then be written as:

$$X = (s, y, \tilde{\theta})^T \tag{1}$$

Finally, since a kinematic model is here investigated, vehicle *control vector* is:

$$U = (v, \delta)^T \tag{2}$$

Vehicle state space model is now derived, considering successively two possible wheels-ground contact conditions.

3.2. Vehicle Modeling under Non-sliding Assumption

In this section, standard pure rolling and non-sliding contact conditions are assumed to be satisfied. They imply that the linear velocity vector at each wheel center belongs to the wheel plane. When applied to the bicycle model shown on Figure 4, one can obtain that the linear velocity vector at the virtual front wheel center presents an angle δ with respect to the vehicle centerline, and the linear velocity vector at the virtual rear wheel center O, previously denoted v, is directed along the vehicle centerline.

Now, just remind that the vehicle is considered as a rigid body. This ensures that, at each instant, its motion is either a pure translation, or a pure rotation around a moving point termed *Instantaneous Rotation Center (IRC)*. Actually, both situations can be gathered into only one, since translations are just special rotations for which the IRC has moved to infinity. This IRC is clearly defined as the intersection point of the perpendiculars to the linear velocity vectors at any 2 points of the rigid body. This is obvious in the pure rotation case. In the pure translation case, the linear velocity vectors are parallel, the intersection point of their perpendiculars is then consistently rejected to infinity.

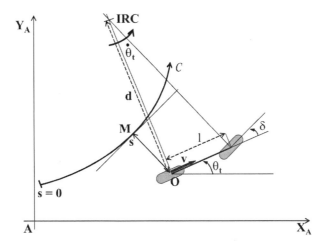

Figure 5. Derivation of the $\dot{\theta}_t$ angular velocity equation.

On Figure 5, the IRC location is drawn from the perpendiculars to the linear velocity vectors at point O and at the front wheel center, whose directions have been pointed out in the above paragraph. Let us finally denote d, the distance between the IRC and O. The vehicle absolute angular velocity $\dot{\theta}_t$ can then easily be derived: relying on the celebrated relation between angular and linear velocities, it can be obtained that (see Figure 5):

$$\dot{\theta}_t = \frac{v}{d} \qquad (3)$$

The value of d can be easily inferred from basic geometrical relations:

$$\tan \delta = \frac{l}{d} \qquad (4)$$

Therefore, gathering (3) with (4) provides us finally with:

$$\dot{\theta}_t = \frac{v}{l} \tan \delta \qquad (5)$$

Let us now address the derivation of $\dot{\theta}_c$, relying on Figure 6, where $R(s)$ denotes the curvature center of path $\mathcal{C}$ at the curvilinear coordinate s.

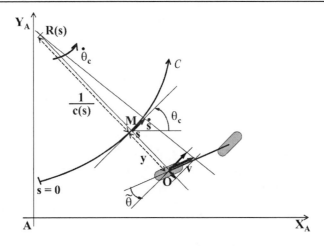

Figure 6. Derivation of the $\dot{\theta}_c$ angular velocity equation.

By definition, the distance between $R(s)$ and M is $\frac{1}{c(s)}$. Using again the relation between angular and linear velocities, it follows that (see Figure 6):

$$\dot{\theta}_c = \frac{\dot{s}}{\frac{1}{c(s)}} = \frac{v \cos \tilde{\theta}}{\frac{1}{c(s)} - y} \qquad (6)$$

Note that y is negative on Figure 6. This explains the minus sign in equation (6). It can be deduced immediately from this latter relation that:

$$\dot{s} = \frac{v \cos \tilde{\theta}}{1 - y\, c(s)} \qquad (7)$$

$$\dot{\theta}_c = \frac{c(s)\, v \cos \tilde{\theta}}{1 - y\, c(s)} \qquad (8)$$

It is also immediate from Figure 6 that:

$$\dot{y} = v \sin \tilde{\theta} \qquad (9)$$

Note that $\tilde{\theta}$ is negative on Figure 6. Signs are then consistent in equation (9). Gathering relations (5), (7), (8) and (9) provides us finally with the *vehicle kinematic model when wheels-ground contact conditions satisfy pure rolling and non-sliding assumptions*:

$$\begin{cases} \dot{s} = v \frac{\cos \tilde{\theta}}{1 - y\, c(s)} \\ \dot{y} = v \sin \tilde{\theta} \\ \dot{\tilde{\theta}} = v \left(\frac{\tan \delta}{l} - \frac{c(s) \cos \tilde{\theta}}{1 - y\, c(s)} \right) \end{cases} \qquad (10)$$

It can be noticed, that Model (10) becomes singular when $y = \frac{1}{c(s)}$, i.e. when points O and $R(s)$ are superposed. This problem is not encountered in practical situations: on one hand, path curvatures are always small, and on the other hand, it is expected that the farm tractor remains close to C.

3.3. Vehicle Modeling Accounting for Sliding Effects

In order to account for sliding effects, that inevitably occur in agricultural applications, one possibility is to extend standard kinematic Model (10) with an additive structured perturbation. Such an approach has already been satisfactorily applied to nautical applications, in order to account for stream effects on vessels motion, see [7] and [10]. Since stream acts on vessels in a similar way than sliding acts on land vehicles, this approach is here below investigated.

Let us again describe the farm tractor according to a bicycle model. When sliding occurs, the ground reaction is no longer equal to the wheel action on the ground. The resultant forces at both wheel-ground contact points can be decomposed into longitudinal forces, directed along the wheels planes, and lateral forces, perpendicular to these planes. However, since only lateral guidance applications are here addressed (i.e. the control objective is that the vehicles follow reference paths with a high accuracy), longitudinal sliding resultant forces can be disregarded: first, these forces preponderantly influence the vehicle velocity, whose control is not addressed in path following applications (v is a free control variable, to be tuned by the farmer). Moreover, the performances of lateral guidance laws are shown in forthcoming Section 4. to be independent from the vehicle velocity. Therefore longitudinal sliding forces have actually a very weak impact in lateral guidance applications, contrarily to lateral sliding forces, denoted $\vec{F}_{front}$ and $\vec{F}_{rear}$ in Figure 7.

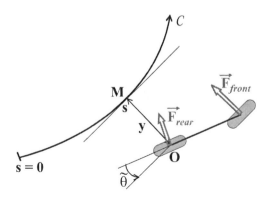

Figure 7. Lateral sliding resultant forces.

These two forces obviously generate:

- *a lateral resultant force*, since the projection of $\vec{F}_{front}$ and $\vec{F}_{rear}$ on straight line (MO) is non-null. This leads the vehicle to move sideways.

- *a resultant torque*, since $\vec{F}_{front}$ and $\vec{F}_{rear}$ are generally not identical (due to mass distribution, tire structure, contact conditions at wheel-ground contact points, etc.). This leads the vehicle to turn on itself.

These effects, deduced from schematic Figure 7, can however be also exhibited via thorough vehicle dynamic analysis, see e.g. [4].

Therefore, a *vehicle kinematic model accounting for sliding effects* can be derived by introducing into standard vehicle Model (10) an additive structured perturbation, constituted of a linear lateral velocity $\dot{Y}_p$ and an angular one $\dot{\Theta}_p$:

$$\begin{cases} \dot{s} &= v \, \frac{\cos\tilde{\theta}}{1-y\,c(s)} \\ \dot{y} &= v\,\sin\tilde{\theta} + \dot{Y}_p \\ \dot{\tilde{\theta}} &= v \left(\frac{\tan\delta}{L} - \frac{c(s)\cos\tilde{\theta}}{1-c(s)\,y} \right) + \dot{\Theta}_p \end{cases} \qquad (11)$$

The capabilities of *extended kinematic Model* (11) in describing the vehicle motion in the presence of sliding are demonstrated via full-scale experiments reported in forthcoming Section 5.. For the sake of shortening further expressions, the vehicle sliding variables are denoted P in the sequel:

$$P = (\dot{Y}_P, \, \dot{\Theta}_P)^T \qquad (12)$$

3.4. Measurement and Estimation of Vehicle Variables

In order that guidance law design could rely on Model (11), all vehicle variables, namely state vector X and perturbation vector P, have to be available online. As it has been described in Section 2., the only exteroceptive sensor embarked on the farm tractor is an RTK GPS. Direct measurement and/or estimation of vehicle variables from the information supplied by this single sensor are discussed here below.

3.4.1. Direct Measurement of the Vehicle Location

Capabilities of RTK GPS sensor have been presented in Section 2.: this sensor can supply the absolute position of its antenna center with a 2 *cm* accuracy at a 10 *Hz* frequency.

The GPS antenna has been located on the top of the tractor cabin, accurately straight up above the point O, see Figure 4. Therefore, if the tractor roll and pitch are zero, the absolute position of point O is available by direct measurement. From the knowledge of reference path $\mathcal{C}$, the location of point M can then easily be inferred. This allows us to have access to the two first coordinates of the tractor state vector: when the tractor roll and pitch are zero, s and y are available online with the same high accuracy than the sensor one.

Since the farm tractors move on irregular ground, roll and pitch are not actually zero. However, in most encountered situations, these two angles present quite small values. Therefore, accuracy of s and y direct measurement is slightly less than the sensor one, but is still satisfactory in order to address accurate automatic guidance applications.

3.4.2. Reconstruction of the Vehicle Heading

When pure rolling and non-sliding conditions are satisfied at wheels-ground contact points, the vehicle linear velocity at point O is directed along the vehicle centerline, as previously mentioned in Section 3.2.. Therefore, if reliable velocity measurements were available, the farm tractor heading θ_t could be obtained straightforwardly: let (v_{x_A}, v_{y_A}) denote the coordinates of the linear velocity vector v in the absolute frame $[A, X_A, Y_A)$. Then, θ_t could be computed online according to (see also Figure 4):

$$\theta_t = \begin{cases} \arctan\frac{v_{y_A}}{v_{x_A}} & \text{if } v_{x_A} \neq 0, \\ \text{sign}(v_{y_A})\frac{\pi}{2} & \text{if } v_{x_A} = 0 \end{cases} \qquad (13)$$

Moreover, since the current value of s is already online available (see above), the desired vehicle heading $\theta_c(s)$ can be inferred from the knowledge of reference path $\mathcal{C}$. Therefore, from a theoretical point of view, the last coordinate of the tractor state vector, namely $\tilde{\theta} = \theta_t - \theta_c(s)$, can then also be obtained online, via direct measurements.

As above mentioned, when the tractor roll and pitch are zero, the single RTK GPS sensor provides us online with the accurate absolute position of point O. The coordinates of velocity vector v can then clearly be inferred, by simply differentiating two successive position information. However, such measurements cannot provide us with accurate enough values of θ_t. In order to illustrate that point, let us consider a farm tractor describing a perfect straight line (i.e. δ is constantly equal to 0) at $8\ km.h^{-1}$. The vehicle actual positions at two successive sample times ($T_s = 100\ ms$, in view of the GPS sensor sampling frequency) are depicted on Figure 8. When the tractor roll and pitch are zero, the vehicle positions provided by the GPS sensor, in view of its accuracy, are inside a 2 cm radius circle centered on the actual point O position. Therefore in the worst case, the vehicle heading computed from (13) could be $atan(4/22) = 10.3°$ instead of a null expected heading, as shown on Figure 8. This theoretical maximum heading measurement error is of course larger if the vehicle velocity is smaller. Moreover, if the tractor roll and pitch are not zero and/or the vehicle undergoes sliding, the accuracy of vehicle heading direct measurement may still be significantly damaged.

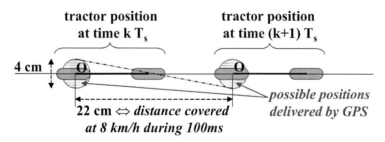

Figure 8. Theoretical computation of the accuracy of the heading direct measurement.

Straight line runs at $8\ km.h^{-1}$ have been performed with our experimental set up, in order to investigate whether the theoretical maximum heading measurement error computed above is representative for actual applications or not. Experimental data reveal that θ_t maximum value provided by relation (13) is $11.81°$ (instead of a null expected heading) and that the standard deviation of θ_t values is $2.4°$. These experimental data are therefore consistent with theoretical computations depicted on Figure 8.

If θ_t direct measurements were used straightforwardly inside automatic guidance law (designed in forthcoming Section 4.), computed steering angle δ would clearly be very oscillating. Depending on the actuator bandwidth, these oscillations could either be transmitted to the front wheels or be filtered out. In the latter situation, which is however not at all a very satisfactory one, the oscillations in θ_t direct measurements would not be perceptible on the vehicle motion. In order to settle this point, spectral analysis has been carried out on experimental data recorded in the above mentioned experiments. The spectral modes of θ_t signal provided by direct measurements (i.e. computed from RTK GPS information and relation (13)) are shown in dash-dotted line on Figure 9. During these experiments, a vehicle heading measurement device (consisting in a double GPS antenna) had also exceptionally been embarked on the tractor, in order to provide with more reliable θ_t values. The spectral modes of this latter signal are shown in dashed line on Figure 9.

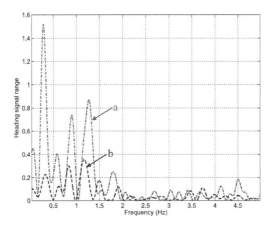

Figure 9. Spectral analysis of θ_t direct measurement (dash-dotted - a) and θ_t provided by a double GPS antenna (dashed - b).

It can be observed on Figure 9 that the θ_t signal delivered by a single RTK GPS according to (13) displays very low frequency modes (0.3 Hz, 1 Hz, ...). These modes could definitely not be filtered out by the steering actuator, since its bandwidth can be shown to be close to 2 Hz, see forthcoming Section 4.3.. With a much lower order of magnitude, the same low frequency modes are observed on the θ_t signal delivered by the double GPS antenna device. Therefore, it appears that the oscillations in the θ_t signal delivered by a single RTK GPS originate from perturbations such as the tractor roll and pitch and/or sliding effects, and then are amplified since velocity measurements cannot be provided with a sufficient accuracy, as highlighted on Figure 8.

The spectral analysis depicted on Figure 9 demonstrates that, if automatic guidance is intended to be achieved relying solely on a RTK GPS sensor, then the vehicle heading θ_t must be filtered prior to be sent to the steering law. Since tractor models are available, an efficient alternative to standard digital filters consists in making use of them through a *Kalman state reconstructor*.

Let us assume again that pure rolling and non-sliding conditions are satisfied at wheels-ground contact points. Then, θ_t equation is given by (5). Since v and δ are both control variables, this equation is a priori a non-linear one. However:

- on one hand, for obvious practical reasons, δ is bounded:

$$|\delta| < \delta_{max} < \frac{\pi}{2}$$

Therefore, $\tan \delta$ can be regarded as a control variable in the place of δ.

- on the other hand, since the control objective in forthcoming Section 4. is to achieve path following, v is not actually regarded as a control variable: it is a free parameter to be tuned by the farmer, whose value may possibly be slowly varying.

In view of these two remarks, equation (5) can actually be considered as a linear equation, and celebrated Kalman linear state reconstructor can be used. More precisely, discrete analogue of equation (5) is:

$$\theta_{t,[k]} = \theta_{t,[k-1]} + \frac{v\,T_s}{l}\tan\delta_{[k-1]} \tag{14}$$

where T_s is the sampling period. The model and the innovation equations of the Kalman state reconstructor associated with Model (14) are, see e.g. [6]:

$$\begin{cases} \bar{\theta}_{t,[k]} &= \hat{\theta}_{t,[k-1]} + \frac{v\,T_s}{l}\tan\delta_{[k-1]} \\ \hat{\theta}_{t,[k]} &= \bar{\theta}_{t,[k]} + L\left(\theta_{t,[k]} - \bar{\theta}_{t,[k]}\right) \end{cases} \tag{15}$$

$\theta_{t,[k]}$ is the kth sample of the raw vehicle heading θ_t, derived from the velocity measurements supplied by the RTK GPS and relation (13). $\bar{\theta}_{t,[k]}$ is the kth prediction of signal θ_t, and $\hat{\theta}_{t,[k]}$ is the kth sample of the filtered heading signal, that will be used in the forthcoming guidance laws. Finally, L is the scalar Kalman gain, to be chosen with respect to the RTK GPS sensor features. In all forthcoming experiments, L has been tuned to 0.08.

The capabilities provided by heading reconstructor (15) have been investigated via two sets of experiments.

First, straight line runs have again been performed. The vehicle heading θ_t, which is expected to be constantly zero, has been computed from the velocity measurements supplied by the RTK GPS and relation (13), and then filtered according to (15). Filtering benefits appear clearly in Table 2:

Table 2. θ_t measurement / reconstruction during straight line runs

	max. value	standard deviation
θ_t computed directly from (13)	11.81°	2.4°
θ_t computed from reconstructor (15)	3.61°	0.86°

In view of Table 2, in a steady state (i.e. when the vehicles describe straight lines), accuracy of the vehicle heading θ_t provided by Kalman reconstructor (15) appears satisfactory and consistent with path following applications.

The capabilities of reconstructor (15) when the vehicle heading is varying, and especially the delay introduced by that filter, have been investigated via a second set of experiments: the farm tractor has been manually driven in order to describe successively a straight line, then the quarter of a circle, and finally a straight line. The vehicle heading θ_t, either computed directly from the velocity measurements provided by the RTK GPS and relation (13), or proceeded through Kalman reconstructor (15), is displayed on Figure 10 respectively in dash-dotted and solid lines. These signals are compared with the signal supplied by a reliable heading measurement device (namely a double GPS antenna) shown in dashed line on Figure 10.

It can be seen on Figure 10 that the vehicle heading θ_t, when computed directly from the velocity measurements supplied by the RTK GPS and relation (13), is a very noisy signal.

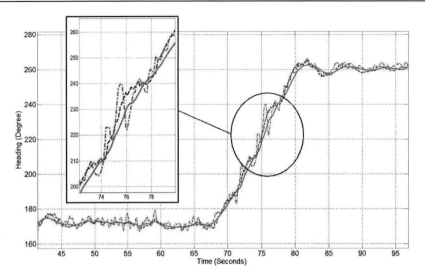

Figure 10. Vehicle heading θ_t signals. *Dashed line*: double GPS antenna, *dash-dotted line*: computed directly from (13), *solid line*: obtained from reconstructor (15).

This negative feature is even more important when the vehicle heading is varying, see the zoom shown on Figure 10. The guidance laws to be designed in forthcoming Section 4. cannot clearly rely on a so noisy heading information. In contrast, it can be observed on Figure 10 that the vehicle heading θ_t provided by Kalman reconstructor (15) follows very closely the reliable vehicle heading supplied by the double GPS antenna. More precisely, it can be noticed that:

- as expected, the vehicle heading θ_t proceeded through Kalman reconstructor is no longer a noisy signal,

- moreover, reconstructor equations (15) have been derived under assumptions of pure rolling and non-sliding conditions at wheels-ground contact points. When the farm tractor enters into the circular part of its path, these assumptions are no longer satisfied: sliding necessarily occurs. Nevertheless, the vehicle heading θ_t derived from equations (15) still fits quite satisfactorily with the actual vehicle heading supplied by the double GPS antenna.

- the delay introduced by the Kalman reconstructor, even when sliding occurs, never climbs over 300 *ms*, see the zoom shown on Figure 10. This is not prohibitive with respect to the considered applications (heavy farm tractors cannot change significantly their heading in a so short time).

Therefore, embarking an additional heading measurement device (such as a double GPS antenna) does not appear mandatory in order to achieve path following. In all guidance experiments reported in forthcoming Section 5., the vehicle heading will be provided from the information supplied by the single RTK GPS sensor, and proceeded through Kalman reconstructor (15).

3.4.3. Estimation of the Sliding Variables

As shown above, the single RTK GPS sensor can provide online with the whole vehicle state vector X. Therefore, it enables to achieve path following, as long as sliding effects have not to be accounted. Such path following control laws, relying on Model (10) (i.e. model without sliding), are designed in Section 4.1., and related full-scale experiments are reported in Section 5.1.. As expected, satisfactory guidance accuracy is displayed as long as the vehicles do not enter into sharp curves or move on sloping fields. In these two latter situations, sliding occurs and decreases guidance accuracy.

In order to preserve the guidance accuracy in these situations, sliding effects have to be accounted explicitly in control laws. This imposes that perturbation vector P must also be available online. A direct measurement would require sophisticated devices, and their accuracies would not be guaranteed on irregular agricultural fields. Therefore, it is proposed here to still rely on the single RTK GPS sensor, and to estimate P according to an *Internal Model Adaptive scheme*.

More precisely, at each sample time, the RTK GPS sensor provides with the actual vehicle behavior, assumed to obey Model (11). Perturbation vector P can then be derived by comparing this actual behavior to the theoretical behavior in the absence of sliding, which can be obtained by simulating Model (10) with the actual steering angle δ applied to the vehicle. This estimation scheme is depicted on Figure 11.

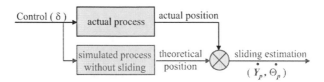

Figure 11. Sliding variables estimation scheme.

Explicit estimation equations can be derived from Models (10)-(11) and relation (5):

$$\begin{cases} \dot{Y}_{P,[k]} = \frac{y_{[k]} - y_{[k-1]}}{T_s} - v \sin \tilde{\theta}_{[k-1]} \\ \dot{\Theta}_{P,[k]} = \frac{\theta_{t,[k]} - \theta_{t,[k-1]}}{T_s} - v \frac{\tan \delta_{[k-1]}}{L} \end{cases} \quad (16)$$

As usual, T_s denotes the sampling period and subscripts $_{[k]}$ refer to the kth sample. It is worth noticing that perturbation vector P is not computed from absolute lateral and angular deviations, but more conveniently from relative deviations occurring during a sampling period. Sliding estimation algorithm is therefore independent from vehicle absolute deviation.

It is clear from equations (16) that perturbation vector P does not account specifically for sliding effects: any perturbation acting on standard Model (10) in a similar way than sliding effects (i.e. by introducing additional lateral and angular velocities) is also accounted. Therefore, the values presented by perturbation vector P are mainly explained by sliding occurrence, but other dynamic effects may also be superposed inside P: delays originating from vehicle large inertia, oscillations due to vehicle roll and pitch, etc. The aggregation of the perturbations acting on the vehicle exhibits positive and negative features:

- <u>positive feature</u>: in forthcoming Section 4.2., vehicle guidance laws are modified, according to adaptive control techniques, in order to annihilate the impact of perturbation vector P on guidance accuracy. The main objective of this adaptive control design is to preserve guidance accuracy when the vehicles undergo sliding. However, the same positive action is obtained with respect to any other perturbation accounted in P. Therefore, vehicle guidance accuracy is finally also improved with respect to some other dynamic phenomena that had not been explicitly considered when designing Model (11) (such as e.g. delays originating from the vehicle large inertia).

- <u>negative feature</u>: since perturbation vector P does not specifically describe sliding effects and some well-identified dynamic phenomena, it will be a quite noisy signal. For instance, due to ground irregularities, the tractor cabin undergoes necessarily some roll. Since the GPS antenna is located on the cabin, roll alters lateral deviation measurements and, in view of equations (16), roll is finally incorporated into perturbation vector P, despite it has no effect on point O location. Moreover, since measurement data are somehow differentiated in estimation equations (16), these parasitic phenomena could be significantly amplified.

Therefore, perturbation vector P must be filtered in order to preserve the useful information and, at minimum, be accurately representative for sliding effects, as desired. Since no model is available for parasitic phenomena, standard digital filters have been investigated: more precisely, a first order Butterworth low-pass filter has been considered, in order to avoid the introduction of too much delay in sliding estimation.

The capabilities of Model (11) and estimation algorithm (16) to account for sliding effects in the vehicle motion have been investigated via full-scale experiments. Satisfactory results, recorded when the farm tractor describes sharp curves on a level ground, or when it describes straight lines on a sloping field, are reported and discussed in forthcoming Section 5.. Control law design, relying on Model (11), is now addressed in the next Section, with the aim to achieve accurate path following, even when the vehicles undergo sliding.

4. Path Following Control Law Design

The control objective is to provide farm vehicles with accurate path following capabilities. The reference path is known: it has either been computed from a GPS map of the field, or been experimentally learnt (i.e. the farm tractor is manually driven along the desired reference path, and this latter is then recorded from the tractor RTK GPS sensor).

Vehicle modeling presented in Section 3. is very suitable to address this problem, since the control objective can be expressed very simply as *"bringing and keeping lateral deviation y to 0"*. When the vehicles do not undergo sliding, angular deviation $\tilde{\theta}$ is also expected to converge to 0 (this intuitive feature can also be recovered from the second equation in Model (10)). In contrast, when sliding occurs, the convergence of y is no longer consistent with that of $\tilde{\theta}$: it can be derived from the second equation in Model (11) that the asymptotic value of $\tilde{\theta}$ depends on $\dot{Y}_P$. From a practical point of view, when sliding occurs, the vehicles have to move slightly crabwise in order to ensure a null lateral deviation.

Path following applications are not concerned with vehicle velocity control: v must be a free parameter that can be tuned online by the farmer according to the agricultural task to be achieved. Therefore, with respect to guidance control laws, v has to be considered as a parameter whose value may possibly be slowly varying. The sole available control variable is the steering angle δ.

This section is organized as follows: first, vehicle automatic guidance when sliding does not occur is addressed in Section 4.1.. Non-linear control techniques are applied on Model (10), and provide with very satisfactory guidance performances. In order to preserve these performances when sliding occurs, previous guidance laws are refined in Section 4.2., relying on adaptive control techniques applied from Model (11). Guidance performances are shown to be satisfactory excepted at sliding appearance/disappearance. The overshoots that can be observed originate mainly from the delay introduced by the steering actuator. Model predictive control techniques are then investigated in Section 4.3. to cope with this difficulty. Performances of the proposed guidance laws are finally displayed from numerous full-scale experiments reported in forthcoming Section 5..

4.1. Non-linear Control in Absence of Sliding

When sliding effects are not accounted, the vehicle motion is described by Model (10). This model is clearly a non-linear one. In [1], the authors propose to linearize it around the equilibrium $y = \tilde{\theta} = 0$, in order that celebrated Linear Systems theory could be used. In that case, control design does not rely on the actual vehicle model, but on an approximated one. Recent advances in Control theory have established that mobile robots models can be converted into almost linear models, namely *chained forms,* in an exact way, see e.g. [17]. Such an approach is attractive since it allows to use, for a large part, Linear Systems theory, while still relying on the actual non-linear vehicle model. This approach is followed in the sequel.

4.1.1. Conversion of Vehicle Model (10) into Chained Form

The general chained form dedicated to systems with two inputs is written as (see [17]):

$$
\left\{
\begin{array}{rcl}
\dot{a}_1 & = & m_1 \\
\dot{a}_2 & = & a_3\, m_1 \\
\dot{a}_3 & = & a_4\, m_1 \\
\cdots & & \cdots \\
\dot{a}_{n-1} & = & a_n\, m_1 \\
\dot{a}_n & = & m_2
\end{array}
\right.
\tag{17}
$$

with $A = (a_1, a_2, \ldots, a_n)^T$ and $M = (m_1, m_2)^T$ respectively the state and control vectors. In order to point out that a chained system is almost linear, just replace the time derivative by a derivation with respect to the state variable a_1. Using the notations:

$$
\frac{d}{d\,a_1} a_i = a_i' \qquad \text{and} \qquad m_3 = \frac{m_2}{m_1}
\tag{18}
$$

the chained form (17) can be rewritten:

$$\begin{cases} a'_1 &=& 1 \\ a'_2 &=& a_3 \\ a'_3 &=& a_4 \\ \dots & & \dots \\ a'_{n-1} &=& a_n \\ a'_n &=& m_3 \end{cases} \tag{19}$$

The last $n-1$ equations of system (19) constitute clearly a linear system.

Let us now convert vehicle Model (10) into chained form. When limited to dimension 3, the general chain systems (17) and (19) are written respectively as:

$$\text{derivation w.r. to time}: \quad \begin{cases} \dot{a}_1 &=& m_1 \\ \dot{a}_2 &=& a_3\, m_1 \\ \dot{a}_3 &=& m_2 \end{cases} \tag{20}$$

$$\text{derivation w.r. to } a_1: \quad \begin{cases} a'_1 &=& 1 \\ a'_2 &=& a_3 \\ a'_3 &=& m_3 \end{cases} \tag{21}$$

Since control law performances are expected to be independent from the vehicle velocity, the variable a_1, which drives the linear system (21), should be homogeneous at the distance covered by the vehicle. A natural choice is then:

$$a_1 = s \tag{22}$$

Straightforward computations show now that the non-linear vehicle Model (10) can actually be converted into chained forms (20) or (21) from the starting choice (22).

In order to fit with (20), the new control m_1 is necessarily defined as:

$$m_1 \triangleq \dot{a}_1 = v\frac{\cos\tilde{\theta}}{1 - y\,c(s)} \tag{23}$$

Moreover, for the sake of simplicity, let us try:

$$a_2 = y$$

It follows that:

$$\dot{a}_2 = v\sin\tilde{\theta} \triangleq a_3\, m_1$$

Therefore the last state variable a_3 must be chosen as:

$$a_3 = (1 - y\,c(s))\tan\tilde{\theta}$$

Finally, the last control variable m_2 is necessarily given by:

$$\begin{aligned} m_2 \triangleq \dot{a}_3 = & \frac{d}{dt}((1 - y\,c(s))\tan\tilde{\theta}) \\ = & -c(s)\,v\sin\tilde{\theta}\tan\tilde{\theta} - \frac{d\,c(s)}{d\,s}\frac{v\cos\tilde{\theta}}{1 - y\,c(s)}\tan\tilde{\theta}\,y \\ & +v\frac{(1 - y\,c(s))}{\cos^2\tilde{\theta}}\left(\frac{\tan\delta}{l} - c(s)\frac{\cos\tilde{\theta}}{1 - y\,c(s)}\right) \end{aligned} \tag{24}$$

As a conclusion, the non-linear tractor Model (10) can be converted into chained forms (20) or (21) in an exact way according to the state transformation:

$$A = \Psi(X) \quad \text{with} \quad \Psi(X) = (s, y, (1 - y\,c(s))\tan\tilde\theta)^T \tag{25}$$

and the control transformation:

$$M = \Upsilon(U,\,X) \text{ defined by (23) and (24)} \tag{26}$$

These transformations are invertible as long as $y \neq \frac{1}{c(s)}$ (model singularity), $v \neq 0$, and $\tilde\theta \neq \frac{\pi}{2}\,[\pi]$. From a practical point of view, once properly initialized, the guided vehicle respects these conditions.

4.1.2. Non-linear Control Law Design

Control design can now be completed in a very simple way: since chained form (21) is linear, a natural expression for the virtual control law is:

$$m_3 = -K_d\,a_3 - K_p\,a_2 \quad (K_p, K_d) \in \mathcal{R}^{+2} \tag{27}$$

As a matter of fact, injecting (27) into (21) leads to:

$$a_2'' + K_d\,a_2' + K_p\,a_2 = 0 \tag{28}$$

which implies that a_2 converges to zero. In view of (25), the convergence of y to 0 is therefore ensured: path following is clearly achieved. Moreover, the convergence of a_2 implies that of a_3 (in view of chained form (21)). State transformation (25) ensures then that angular deviation $\tilde\theta$ converges also to 0.

Since the error dynamics (28) is driven by $a_1 = s$, the gains $(K_d,\,K_p)$ impose a settling distance instead of a settling time. Consequently, for a given initial error, the vehicle trajectory will be identical, whatever the value of v is, and even if v is time-varying. From a control design point of view, guidance performances are velocity independent: control law gains have not to be adjusted with respect to vehicle velocity v. In practical situations, this theoretical result might be slightly altered, since the quality of θ_t (and therefore of $\tilde\theta$) measurement clearly depends on v (relations (13) and (15)), or since the tractor actuators are not perfectly linear. Nevertheless, as long as standard agricultural velocities (from 4 to 14 $km.h^{-1}$) are concerned, and provided that control gains $(K_d,\,K_p)$ are not so high that actuators are saturating, experimental results demonstrate that guidance performances are actually velocity independent, see forthcoming Section 5.1..

Ultimately, the inversion of control transformations (26) provides with the actual control law expression (just report (27) in (18), (23) and (24)):

$$\begin{aligned}
\delta(y, \tilde\theta) = \arctan\Big(l\Big[&\frac{\cos^3\tilde\theta}{(1 - y\,c(s))^2}\Big(\frac{d\,c(s)}{d\,s}\Big)y\tan\tilde\theta \\
&- K_d\,(1 - y\,c(s))\tan\tilde\theta - K_p\,y \\
&+ c(s)\,(1 - y\,c(s))\tan^2\tilde\theta\Big) + \frac{c(s)\,\cos\tilde\theta}{1 - y\,c(s)}\Big]\Big)
\end{aligned} \tag{29}$$

It is worth noticing that no approximation has been introduced in control design. This enables non-linear control law (29) to ensure accurate curved path following: non-linearities of Model (10) and curvature $c(s)$ of the reference path are both explicitly accounted in control expression. In addition, performances tuning of non-linear control law (29) is still very easy, since it relies on linear error equation (28): a settling distance can very intuitively be imposed by just adjusting gains K_p and K_d. Moreover, since a settling distance instead of a settling time is imposed, guidance law performances are actually independent from the vehicle velocity, even if v is varying.

In many applications, the reference path $\mathcal{C}$ is a straight line, i.e. $c(s) = 0$. The expression of the control law (29) turns then simpler:

$$\delta(y, \tilde{\theta}) = \arctan(l \cos^3 \tilde{\theta} (-K_d \tan \tilde{\theta} - K_p y)) \tag{30}$$

Finally, let us go back to the discussion on actuators saturation. In control laws (29) or (30), the argument of the *arctan* function is not bounded. Therefore actuators saturation can a priori occur. The natural way to deal with it, is to adjust control performances (i.e. to tune gains (K_d, K_p)) in order that saturations are never met during prespecified operations. However, it can be pointed out that actuators saturation does not prevent from the vehicle convergence to the reference path $\mathcal{C}$, even from a theoretical point of view: since chained form (21) consists in a double integrator, its asymptotic stability is still ensured, even if the virtual control law (27) is bounded to any arbitrary value, see [18]. Unfortunately, in view of (29), the boundedness of m_3 leads to that of δ only if the reference path curvature $c(s)$ exhibits some good properties. However, in most practical situations, these properties are satisfied, so that the theoretical stability is actually preserved. For instance, it is obviously checked when $\mathcal{C}$ is a straight line, since $c(s) = 0$. The only drawback is that control performances are, of course, no longer velocity independent as soon as the actuators are saturating.

Actually, the main difficulty proposed by the steering actuator is not saturations, but the introduction of delays into the guidance feedback loop. The influence of these delays is very perceptible when sliding occurs, as it can be seen on experimental reports in Section 5.2.. This actuator feature is accounted in Section 4.3., relying on Model Predictive techniques.

4.2. Internal Model Adaptive Control Accounting for Sliding Effects

Sliding effects are described in Model (11) as an additive structured perturbation. When such additive perturbations are constant or slowly-varying, they can be easily rejected by introducing integral correction terms into control law design. However, in agricultural tasks, sliding is definitely not a static perturbation. In contrast, its dynamics exhibits high frequency modes. Therefore, sliding effects cannot be accounted accurately by simply adding integral correction terms into control law (29). Full-scale experiments have been carried out and have corroborated this conclusion.

Adaptive control techniques propose a more convenient framework to cope with additive structured perturbations. Such an approach has for instance been developed in [3] for a harvesting application relying on a video camera sensor: when sliding effects are detected, a correction term computed from an empirical relation is incorporated into the guidance law previously derived under classical pure rolling and non-sliding contact conditions. Full-scale experiments demonstrate that this adaptive correction scheme improves significantly guidance accuracy in the presence of sliding.

In this Section, *Internal Model adaptive control* techniques, see e.g. [2]-[11], are used to deal with sliding effects. The major advantage of this approach is that it allows to account for sliding while still relying on non-linear control law (29) designed in previous section. Thus, all positive features of this guidance law (accuracy, intuitive performances tuning, performances independent from vehicle velocity, . . .) can be preserved.

As a first step, let us address an academic case where perturbation vector P is constant (for instance, when the vehicle describes a straight line on a field with a perfect constant slope). In such a case, it can be shown that non-linear control law (29), which does not account for sliding effects, leads to asymptotic constant guidance errors. More precisely, from Model (11), it can be obtained immediately that:

$$\tilde{\theta} \stackrel{t \to \infty}{\longrightarrow} -\arcsin\left(\frac{\dot{Y}_p}{v}\right) \tag{31}$$

$$\frac{\tan \delta}{L} \stackrel{t \to \infty}{\longrightarrow} -\frac{\dot{\Theta}_p}{v} + \frac{c(s)\cos\tilde{\theta}}{1 - c(s)y} \tag{32}$$

Then, injecting (32) into (29) shows that:

$$\frac{\cos^3\tilde{\theta}}{(1 - c(s)y)^2}(\alpha y + \beta) \stackrel{t \to \infty}{\longrightarrow} -\frac{\dot{\Theta}_p}{v} \tag{33}$$

where:
$$\begin{aligned} \alpha &= \frac{d\,c(s)}{d\,s}\tan\tilde{\theta} + c(s)\tan\tilde{\theta}\,(K_d - c(s)\tan\tilde{\theta}) - K_p \\ \beta &= \tan\tilde{\theta}\,(c(s)\tan\tilde{\theta} - K_d) \end{aligned}$$

Finally, by neglecting second order term y^2 in (33), it can be proved that:

$$y \stackrel{t \to \infty}{\longrightarrow} -\frac{\beta + \dfrac{\dot{\Theta}_p}{v\cos^3\tilde{\theta}}}{\alpha - \dfrac{2\,c(s)\,\dot{\Theta}_p}{v\cos^3\tilde{\theta}}} \stackrel{\Delta}{=} y_c \tag{34}$$

Provided that reference path curvature $c(s)$ is constant or slowly varying, relations (34) and (31) and finally relation (32) establish that lateral and angular deviations (i.e. y and $\tilde{\theta}$) as well as control variable δ asymptotically converge to constant or slowly varying non-null values. This shows that, in response to constant sliding conditions, non-linear control law (29) leads the vehicles to move crabwise. This is somehow consistent with the behavior that can be observed when the vehicles are manually driven.

In the special case where $c(s)$ is constant, the convergence of lateral deviation y to 0, despite the presence of sliding, can be very simply achieved by just shifting the objective of non-linear control law (29): if the objective of control law (29) (which does not account for sliding) is henceforth to bring y to the constant value $-y_c$, then it is clear that the vehicle actual lateral deviation (which undergoes sliding effects) converges to 0. Guidance accuracy is then preserved, although the vehicle is still moving crabwise (relations (31)-(32) are not affected by the control law modification). The expression of the new control law can be deduced straightforwardly from (29):

$$
\begin{aligned}
\delta \;=\; & \arctan \Big(L \Big[\tfrac{cos^3\tilde{\theta}}{(1-c(s)(y+y_c))^2} \Big(\tfrac{d\,c(s)}{d\,s}(y+y_c)\tan\tilde{\theta} \\
& -K_d(1-c(s)(y+y_c))\tan\tilde{\theta} - K_p(y+y_c) \\
& +c(s)(1-c(s)(y+y_c))\tan^2\tilde{\theta}\Big) + \tfrac{c(s)\cos\tilde{\theta}}{1-c(s)y}\Big]\Big)
\end{aligned}
\tag{35}
$$

In agricultural tasks, perturbation vector P and path curvature $c(s)$ are obviously not constant. However, if the shift y_c in the control law objective could track the current sliding conditions, then adaptive non-linear control law (35) would still improve guidance accuracy. Two approaches are proposed below to achieve such an online y_c adaptation:

- *Model Reference Adaptive Control (MRAC)*:

  MRAC scheme is depicted on Figure 12(a). It consists in a simulation application that is constantly running when the vehicle is moving. The equations that are simulated are Model (11) (i.e. vehicle model accounting for sliding). Simulated steering variable δ is computed from control law (29) (i.e. guidance law without sliding accounted), and components of perturbation vector P used in the simulation are the sliding effects acting on the actual vehicle, provided online by estimation algorithm (16). State variable y of this simulation application is then used as the corrective term y_c to be introduced into the control law (35) that steers the actual vehicle.

  When perturbation vector P is constant, the simulated state variable y clearly converges to the theoretical value (34), which is suitable to ensure that control law (35) brings the vehicle lateral deviation to 0, despite the presence of sliding. In the general case, where P is varying, the simulated state variable y is only reflecting current sliding conditions. However, its introduction into control law (35) can intuitively improve path following accuracy. This is corroborated by full-scale experiments reported in Section 5.2..

- *Internal Model Control (IMC)*:

  IMC scheme is depicted on Figure 12(b). In this second approach, the corrective shift y_c to be introduced into control law (35) is directly computed by injecting the actual sliding conditions provided by online estimation algorithm (16) into relation (34).
  When perturbation vector P is constant, the corrective shift y_c is then immediately the suitable one, that can ensure the convergence to 0 of the vehicle lateral deviation. When P is varying, the computed y_c value is also reflecting current sliding conditions, and can also intuitively improve path following accuracy, as corroborated by full-scale experiments reported in Section 5.2..

When perturbation vector P is constant, IMC scheme offers immediately the suitable y_c value. In contrast, this suitable value is not instantaneously available in the MRAC scheme, since one has to wait for the convergence of the simulation application. Therefore, IMC scheme may appear more relevant. However, perturbation vector P delivered by estimation algorithm (16) is a quite noisy signal (as mentioned in Section 3.4.1.). In IMC scheme, this noisy feature is entirely propagated to the corrective shift y_c, since it is directly computed

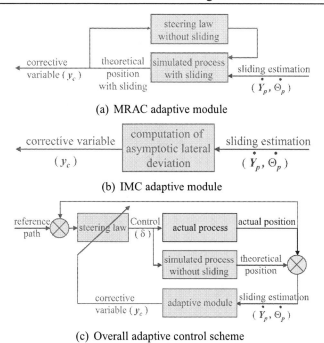

Figure 12. Block-diagrams describing adaptive control scheme.

from the values of P (via relation (34)). Therefore, in order to avoid the transmission of high frequency modes to the steering actuator, y_c values computed from IMC scheme have eventually to be filtered. In contrast, MRAC scheme does not propose the same difficulty since the simulation application acts as a natural low-pass filter. Therefore, none of these two approaches appears clearly superior to the other one.

The overall adaptive control scheme is depicted on Figure 12(c), where the adaptive module consists either in MRAC scheme or in IMC scheme. Nevertheless, whatever this adaptive module is, the proposed guidance law fits with standard Internal Model adaptive control scheme.

4.3. Model Predictive Control Accounting for Actuator Features

The steering actuator embarked on the farm tractor, introduced briefly in Section 2., is an electro-hydraulic valve. As already mentioned, the main limitation of that device is that it introduces an undesired delay in the global guidance feedback loop. The influence of that delay is particularly perceptible when the vehicles undergo sliding: as long as sliding conditions are slowly varying, e.g. when the vehicles are describing a curve, control law (35) satisfactorily ensures an accurate path following. In contrast, when sliding conditions are changing, e.g. when the vehicles enter into a curve or exit from a curve, prohibitive transient guidance errors can be observed. The delay in the actuation, if not the sole responsible, is nevertheless the main responsible for such overshoots. Since this delay is not inherent to our steering actuator, but is more or less present in any actuation device, *Model Predictive control* has been investigated with the aim to bring some anticipation to control law (35), in order to significantly reduce these disappointing overshoots. Moreover, when tuning the predictive parameters, it can be accounted in control law (35), not only for actuation delay, but also for some other dynamic phenomena, such as the vehicle large inertia.

4.3.1. Identification of Steering Actuator Model

In order to support predictive control design, a model describing actuator behavior must be available. This actuator consists in an electro-hydraulic valve controlled via an inner closed-loop scheme, as depicted on Figure 13. The actual front wheel angle δ_a is measured via an absolute encoder and compared with the desired one, denoted δ_d, provided by the guidance law (i.e. control law (29) or (35)). A Proportional-Derivative algorithm, implemented on a PCB80C552 microprocessor, controls then the voltage u applied between the two electrical wires of the valve, in order to adjust the oil flow p such that the front wheels angle δ_a fits at best with δ_d.

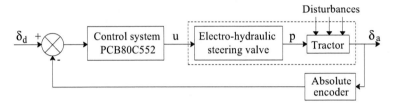

Figure 13. Block-diagram of the steering actuator.

In order to derive the actuator model, several step inputs in desired steering angle δ_d have been applied. The actual steering angle δ_a has then been recorded. It can be obtained from these experiments that the actuation device presents:

- perfect steady state capabilities: the steady state error between δ_d and δ_a is less than 0.1°,

- a 600 *ms* rise time (by the way, this establishes that the actuator bandwidth, questioned in Section 3.4., is actually close to 2 *Hz*)

- a 10% overshoot in step responses.

The 600 *ms* rise time cannot be reduced since it is inherent to the actuator technology, and more precisely to the hydraulic pump capabilities: at tractor nominal engine speed, i.e. 2000 $tr.mn^{-1}$, the maximum oil flow p delivered by the hydraulic pump is limited to 25 $l.mn^{-1}$. At low engine speed (never used in forthcoming experiments), the maximum value of p would even fall to 12 $l.mn^{-1}$, which would still increase inner closed-loop delays.

Relying on these experimental step responses, standard identification techniques establish that this steering actuator can be satisfactorily described by a linear second order model. Its discrete transfert function, to be used in forthcoming predictive control design, is given by:

$$F(z^{-1}) = \frac{\delta_a(z^{-1})}{\delta_d(z^{-1})} = \frac{0.1237\, z^{-1} + 0.0934\, z^{-2}}{1 - 1.2155\, z^{-1} + 0.4326\, z^{-2}} \quad \left(\begin{array}{c} \text{sampling period:} \\ T_s = 0.1\, s \end{array} \right) \quad (36)$$

4.3.2. Control Law (35) Rewriting

In order to reduce transient guidance errors, vehicle control law (35) should anticipate steering action, in order to account for the delay introduced by the actuator. Of course, any situation faced by the vehicles cannot be anticipated: for instance ground irregularities, slippery parts of the field, etc., are completely unpredictable. Actually, the only reliable data on which anticipation can be achieved is the reference path curvature. For instance, in order to reduce guidance overshoots observed when the vehicles enter into a curve, control law values delivered to the actuator could be anticipated, in such a way that, when the vehicles actually enter into the curve, the actual steering angle is consistent with the curvature of the reference path, despite actuator delay. Lateral deviation could then stay close to zero during the transition phase.

In order to introduce such an anticipation into guidance law (35), this latter must first be split into two terms: the first one consisting in the contribution of the reference path curvature to the value of δ, and the second one gathering the contributions of deviations y and $\tilde{\theta}$ and perturbation vector P. Model predictive control techniques will then be applied on the first term, but of course not on the second one.

To achieve such a separation, let us first consider the ideal case where the vehicle is perfectly on the path to be followed and sliding does not occur. In such a situation, the vehicle will stay on the reference path if the curvature defined by the vehicle steering angle is equal to the path curvature, i.e. steering angle δ must satisfy:

$$\frac{\tan \delta}{L} = c(s) \tag{37}$$

This geometric condition can be graphically checked on Figure 4. It can also be deduced from control law expression (35): if deviations $(y, \tilde{\theta})$ and sliding parameters $(\dot{Y}_P, \dot{\Theta}_P)$ are set equal to zero, the only term that is non-null inside the *arctan* parenthesis is the last one, and it provides relation (37) as expected.

Since the main contribution of the reference path curvature to the value of δ is provided by the last term in (35), this term is isolated below from the remaining terms in (35), in order to support anticipation in the forthcoming predictive guidance law. More precisely, let us define:

$$\mu = L \frac{c(s) \cos \tilde{\theta}}{1 - c(s)y} \tag{38}$$

$$\begin{aligned}
\nu = L \Big[&\frac{\cos^3 \tilde{\theta}}{(1 - c(s)(y + y_c))^2} \left(\frac{d\,c(s)}{d\,s}(y + y_c) \tan \tilde{\theta} \right. \\
&- K_d(1 - c(s)(y + y_c)) \tan \tilde{\theta} - K_p(y + y_c) \\
&+ c(s)(1 - c(s)(y + y_c)) \tan^2 \tilde{\theta} \Big) \Big]
\end{aligned} \tag{39}$$

Control law expression (35) can then be written as:

$$\delta = \arctan(\mu + \nu) \tag{40}$$

Relying now on trigonometric relation:

$$\arctan(a + b) = \arctan(a) + \arctan\left(\frac{b}{1 + ab + a^2} \right) \tag{41}$$

vehicle guidance law can finally be split as:

$$\delta = \delta_{Traj} + \delta_{Deviation} \qquad (42)$$

where:
$$\begin{array}{rcl} \delta_{Traj} & = & \arctan(\mu) \\ \delta_{Deviation} & = & \arctan\left(\frac{\nu}{1+\mu\,\nu+\mu^2}\right) \end{array}$$

The two terms in the new guidance law expression (42) exhibit the expected features:

- δ_{Traj} is the main contribution of the reference path curvature to the value of δ. When deviations and sliding are equal to zero, δ_{Traj} provides with the expected value (37). This term will support forthcoming predictive control design.

- $\delta_{Deviation}$ is the contribution of deviations (y, $\tilde{\theta}$) and perturbation vector P (incorporated into y_c) to the value of δ. It aims at bringing y to 0. Since deviations and perturbations are unpredictable, $\delta_{Deviation}$ will remain unchanged in the forthcoming predictive control design.

4.3.3. Predictive Control Design

Model Predictive control techniques (see e.g. [14], [15]) can now be used to provide some anticipation in steering action, in order to reduce the transient guidance errors originating from actuator features. Beforehand, several conventions and notations have to be introduced:

- the kth sample period is chosen below as the current time: state vector $X_{[k]}$ and perturbation vector $P_{[k]}$ are available, and the objective is to compute the steering angle value $\delta_{[k]}$ to be sent to the steering actuator.

- Prediction horizon H:
  In predictive control approach, the current value of the control variables is computed from the current state and perturbation values, but also from the values that some variables should have in the future. H is the time (with respect to current time) when the values of these variables have to be predicted. H is a constant, and must be chosen with respect to the delay introduced by the actuator.
  It is assumed below that $H = h\,T_s$, where T_s is the sample period, and h is an integer.

- Steering objective δ^{Obj}:
  δ^{Obj} is the expected value for the actual steering angle δ_a at time $(k+h)\,T_s$, if at that time deviations and sliding effects were equal to 0.
  More precisely, from the current values $s_{[k]}$ and $v_{[k]}$, the future $s_{[k+h]}$ value can be inferred, and δ^{Obj} can then be derived from relation (37).

- Steering measurement δ^{Meas}:
  $\delta^{Meas}_{[k]}$ is the contribution of δ_{Traj} to the value of the actual steering angle δ_a at current time $k\,T_s$.
  This variable, required in Model Predictive control approach, proposes here a difficulty, since definitely it cannot be measured: the complete control law $\delta =$

$\delta_{Traj} + \delta_{Deviation}$ is always sent to the steering actuator. Therefore, the absolute encoder shown on Figure 13 cannot of course separate the contribution of the two terms in its single measurement $\delta_{a,[k]}$. In the sequel, $\delta_{[k]}^{Meas}$ has been approximated by:

$$\delta_{[k]}^{Meas} = \delta_{a,[k]} - \delta_{Deviation,[k]} \tag{43}$$

- Steering reference δ^{Ref}:

  $(\delta_{[k]}^{Ref}, \delta_{[k+1]}^{Ref}, \ldots, \delta_{[k+h]}^{Ref})$ is the desired trajectory, defined on the prediction horizon H, for the contribution of δ_{Traj} in the actual steering angle δ_a.

  – the trajectory initial value $\delta_{[k]}^{Ref}$ is the actual contribution of δ_{Traj} in δ_a, such as provided by the encoder, i.e. $\delta_{[k]}^{Ref} = \delta_{[k]}^{Meas}$.

  – the trajectory final value $\delta_{[k+h]}^{Ref}$ is the expected value at the end of the prediction horizon. Relying on the above notations, $\delta_{[k+h]}^{Ref} = \delta^{Obj}$.

  – the trajectory linking these initial and final values has been derived according to a standard discrete first order relation:

  $$\delta_{[k+i]}^{Ref} = \delta^{Obj} - \gamma^i \left(\delta^{Obj} - \delta_{[k]}^{Meas} \right) \qquad i \in \{0, 1, \ldots, h\} \tag{44}$$

  where $\gamma \in\;]0,1[$ is a free parameter that enables to shape the trajectory.

- Steering model output δ_a^{Model}:

  $(\delta_{a,[k]}^{Model}, \delta_{a,[k+1]}^{Model}, \ldots, \delta_{a,[k+h]}^{Model})$ is the output of the steering actuator model (36) when the input is $(\delta_{Traj,[k-2]}, \ldots, \delta_{Traj,[k+h-1]})$.

  – $\delta_{Traj,[k-2]}$ and $\delta_{Traj,[k-1]}$ are the actual δ_{Traj} values that have been used in control law (35) and sent previously to the actuator.

  – $(\delta_{Traj,[k]}, \ldots, \delta_{Traj,[k+h-1]})$ are future δ_{Traj} values that are computed from the predictive algorithm detailed below.

These notations are also depicted on Figure 14.

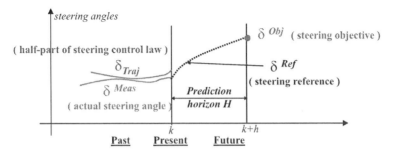

Figure 14. Notations used in Model Predictive Control design.

Model Predictive Control design can now be sketched. Control law (35), previously designed in Section 4.2., provides at each sample period kT_s with the steering angle value

δ that is the most suitable to achieve path following. Unfortunately, this ideal control value is not instantaneously transmitted to the actual front wheels: it is sent to the steering actuation device that will transmit this ideal control value within some delay. Model Predictive Control proposes to anticipate the control value sent to the actuator, relying on the actuator model and on predictions on the future values to be taken by this control variable.

As above mentioned, the only reliable data on which prediction can be achieved is the reference path curvature. Therefore, control law (35) has been split into two terms, δ_{Traj} and $\delta_{Deviation}$ (see expression (42)), and predictive control is now applied only on δ_{Traj}, which encloses the contribution of the reference path curvature to the value of δ. Model Predictive Control consists then in the following steps:

At each sample period $k\,T_s$:

- the value δ^{Obj} that the actual steering angle δ_a is expected to present in the forthcoming H seconds from the current time, is first computed relying on the reference path curvature.

- a trajectory δ^{Ref} which could lead the actual steering angle δ_a from its current value (corrected from $\delta_{Deviation}$, see (43)) to the desired one δ^{Obj} within H seconds is then computed. δ^{Ref} is given by (44).

- for any given future sequence of δ_{Traj}, the future sequence of the actual steering angle can be predicted, relying on actuator Model (36). The sequence $(\delta_{Traj,[k-2]}, \cdots, \delta_{Traj,[k+h-1]})$ that can minimize the deviation between δ^{Model} and δ^{Ref} is then determined from optimization techniques. More precisely, the following problem is solved:

$$seek\,for\,(\,\delta_{Traj,[k-2]}\,,\,\cdots\,,\,\delta_{Traj,[k+h-1]}\,)\ \ such\ that$$
$$J = \sum_{i=0}^{h}\left(\delta^{Model}_{[k+i]} - \delta^{Ref}_{[k+i]}\right)^2\ \ is\ minimum \tag{45}$$

- The value of $\delta_{Traj,[k]}$ obtained from the optimal problem (45), named below $\delta^{Pred}_{Traj,[k]}$, is finally introduced into control law (42).

These computations are iteratively achieved at each further sample period. Therefore, vehicle guidance law accounting for actuator features can finally be written:

$$\delta = \delta^{Pred}_{Traj} + \delta_{Deviation} \tag{46}$$

The overall predictive control law is also depicted on Figure 15.

In order to account properly for actuator features, the optimization of criterion (45) should be performed on a duration superior or equal to the actuator rise time. For the steering device considered here, this imposes $H \geq 0.6\ s$. However, in all experiments, a larger value has been chosen for H, in order to account for steering actuator features, but also for unmodeled dynamic phenomena, such as the large vehicle inertia. These phenomena also introduce delays in the guidance feedback loop, and are therefore also responsible for the transient guidance errors. Enlarging the anticipation provided to δ_{Traj} can obviously reduce their effects. From experimental trials, $H = 1\ s$ has appeared as the most suitable value for the prediction horizon. Finally, γ has been set equal to 0.2 in order to provide δ^{Ref} with a smooth shape.

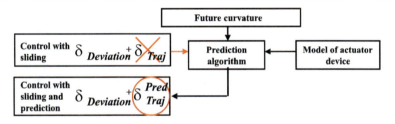

Figure 15. Model Predictive Control scheme.

5. Experimental Results

In order to investigate the capabilities of the guidance laws designed in Section 4., numerous experiments have been carried out in our experimental farm at Montoldre, France.

The first experiments, reported in Section 5.1., have been performed on a level asphalted ground, in order that pure rolling and non-sliding conditions at wheels-ground contact points are satisfied. These experiments aim at highlighting the interesting features of non-linear control law (29), and therefore of control laws (35) and (46), since they are both designed from non-linear control law (29).

The other experiments, reported in Section 5.2., have been performed on agricultural fields. Curved path following on a slippery field, and straight line following on a sloping field are successively discussed. Abilities of control laws (35) and (46) to account respectively for sliding effects and for actuator features are demonstrated.

5.1. Guidance Laws Capabilities in the Absence of Sliding

In this section, it is shown that, as long as the vehicles do not undergo sliding effects, accurate curved path following can be achieved from non-linear control law (29). Additional attractive features of control law (29) are also highlighted.

5.1.1. Step Responses

Although step paths are not common in agriculture, such reference trajectories have first been considered as benchmarks, in order to investigate guidance accuracy when the vehicles describe straight lines and also the convergence features exhibited by control law (29).

Step responses at related velocities

The reference path consists in a 2 meters step. Experiments have been carried out at related velocities, from $v = 2\ km.h^{-1}$ to $v = 14\ km.h^{-1}$ with $2\ km.h^{-1}$ increments. Control law parameters have been tuned according to $(K_d, K_p) = (0.6, 0.09)$. This choice ensures that the error dynamics (28) presents a double pole located at the value 0.3. Linear control tools ensure then that a convergence without any overshoot and within a 15 meters settling distance is specified. This convergence rate has been chosen in order that the tractor motion, especially when its velocity is high, is not so steep that it would be uncomfortable to a person in the tractor cabin. Moreover, it guarantees that actuators saturation is not met. The tractor trajectories are depicted on Figure 16.

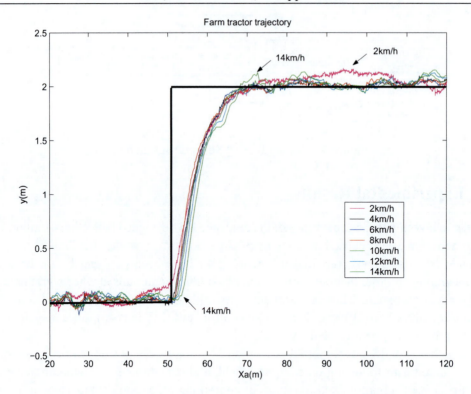

Figure 16. Tractor trajectories during step path following at related velocities.

It can be seen on Figure 16 that all the step responses almost perfectly overlap: as expected, the trajectory is independent from the tractor velocity. Moreover, the 15 meters settling distance, specified when tuning the control parameters, is actually observed. This shows that, although control law (29) is a non-linear one, guidance performances tuning can be achieved in a very simple and intuitive fashion.

A thorough analysis establishes that, excepted for extreme velocities (i.e. $v = 2$ $km.h^{-1}$ and $v = 14\ km.h^{-1}$), path following accuracy is quite satisfactory: once the tractor is following the second straight line (i.e. when $x_A > 70\ m$), the bias λ_y between the tractor trajectory and the reference path, as well as the tractor standard deviation from the mean trajectory σ_y, are both very small:

$$\text{in the worst case:} \quad \lambda_y < 2.7\ cm \quad \sigma_y < 3.1\ cm$$

Finally, one can note that, when the step occurs, the tractor trajectories are more and more shifted when the speed is increased. Obviously, it is a consequence of the delay introduced by the steering actuator: the higher the tractor velocity is, the more the distance covered by the tractor before the steering angle actually changes its value is long.

When the tractor velocity reaches $v = 14\ km.h^{-1}$, the delay introduced by the actuator begins to damage path following performances. It can be observed from Figure 16: on one hand, the tractor starts to react few meters beyond step location, and on the other hand, the tractor trajectory proposes a 10 centimeters overshoot when rejoining the new reference.

$v = 14\ km.h^{-1}$ appears therefore as the maximum tractor velocity compatible with our steering device, when predictive control techniques are not investigated.

When the tractor velocity decreases to $v = 2\ km.h^{-1}$, the overall tractor trajectory is not satisfactory: it can be noticed from Figure 16 that the maximum deviation from the reference path, before or after step location, is beyond 10 centimeters. These very bad performances originate from the vehicle heading measurement. As detailed in Section 3.4., raw vehicle heading θ_t is computed, according to relation (13), from two successive vehicle position measurements provided by the RTK GPS sensor. Although this sensor provides position measurements with a very satisfactory 2 cm accuracy, the accuracy of vehicle heading θ_t is poor: $10.3°$ when $v = 8\ km.h^{-1}$, see Figure 8. Therefore a Kalman state reconstructor (15) has been designed in order to provide control laws with more reliable vehicle heading values. For vehicle velocities that are conventional in agricultural tasks (i.e. $v \approx 10\ km.h^{-1}$), the accuracy of reconstructed vehicle heading is satisfactory, see Table 2. However, the smaller the vehicle velocity is, the worse the accuracy of raw vehicle heading measurements θ_t is: for instance, at very low velocity $v = 2\ km.h^{-1}$, it can be computed from Figure 8 that the theoretical maximum heading measurement error climbs up to $atan(4/5.56)=35.7°$. From such poor raw measurements, Kalman state reconstructor (15) cannot infer accurate vehicle heading values, so that the guidance law performances are necessarily damaged. This is one of the major difficulties that proceed from our initial choice to rely on a *single* RTK GPS receiver. However, it should be emphasized that $v = 2\ km.h^{-1}$ is a very low velocity, which is not typical in agricultural applications.

Step responses with a time-varying velocity

Experiments with a varying tractor velocity have also been carried out: the tractor velocity has been increased linearly from $v = 4\ km.h^{-1}$ to $v = 8\ km.h^{-1}$ when the tractor was performing the step response. Figure 17 displays the tractor trajectory thus obtained (graph c), superposed with two trajectories recorded with a constant tractor velocity (graph a: $4\ km.h^{-1}$, graph b: $8\ km.h^{-1}$)

It can be observed on Figure 17 that all the trajectories are again superposed, and therefore that the 15 meters settling distance is achieved, even when the tractor velocity is varying. This establishes, as expected from the theoretical study presented in section 4.1., that control law (29) is actually velocity independent.

Step responses with large initial conditions

The last step response carried out is reported on Figure 18. It differs from the previous ones, since the initial conditions are very large: initially, the tractor is 10 meters far from the reference trajectory $\mathcal{C}$, and presents a very large heading error $\tilde{\theta} = 65°$. The tractor velocity is $v = 6\ km.h^{-1}$.

Since control law (29) has been designed from the exact non-linear model of the tractor (no approximation, as for instance $\sin \tilde{\theta} \approx \tilde{\theta}$, has been performed), the error dynamics is described *in an exact way* by the linear ODE (28), even if y and $\tilde{\theta}$ are very large. Therefore, it is expected that control law features remain identical to those reported previously.

Figure 18 shows that this theoretical result is actually achieved: it can be noticed from Figure 18(a) that the general appearance of the tractor trajectory is identical to that observed

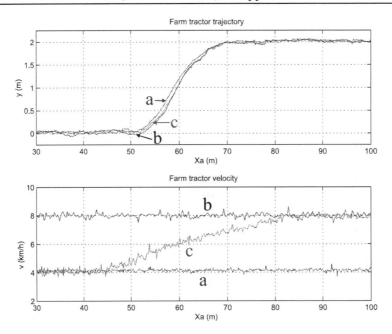

Figure 17. Tractor trajectories during step path following with constant velocities (graph a: 4 $km.h^{-1}$, graph b: 8 $km.h^{-1}$) or a time-varying velocity (graph c).

on Figure 16. Moreover, Figure 18(b) displays that heading deviation $\tilde{\theta}$ is exponentially decreasing within a settling distance equal to 15 meters, as expected.

5.1.2. Sine Curve Following

In order to investigate guidance law (29) capabilities when the reference trajectory is no longer a straight line but a curved path, sine curve following has been achieved. Sine curves are definitely not common trajectories in agricultural tasks. Nevertheless, they are significantly different from straight lines, and therefore can be seen as convincing benchmarks.

The curvature of the sinusoidal reference path has been chosen small (period and peak to peak amplitude are respectively 20 m and 60 cm) such that the vehicles do not undergo sliding when they are guided along this path. Therefore, the capabilities of guidance law (29) with respect to the reference path curvature can here be investigated independently from the incidence of sliding effects.

Experiments have been carried out with control and Kalman parameters K_d, K_p and L identical to those used in previous step response experiments (these parameters are henceforth never changed, even in Section 5.2.). The tractor velocity is $v = 6\ km.h^{-1}$, and the vehicle is initially 60 cm far from the reference trajectory. Its lateral deviation values are reported on Figure 19.

It can be observed on Figure 19 that, as expected, control law (29) ensures the same guidance performances whatever the reference path features are: the settling distance and the statistical variables λ_y and σ_y recorded during this curved path following experiment

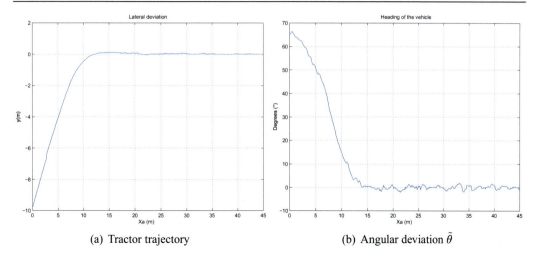

(a) Tractor trajectory   (b) Angular deviation $\tilde{\theta}$

Figure 18. Convergence to a straight line from large initial conditions.

are of the same order of magnitude than those previously recorded during step response experiments.

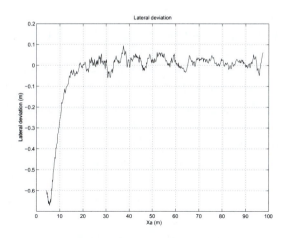

Figure 19. Lateral deviation with respect to a sinusoidal reference path.

Due to numerous occurrences of $c(s)$, the expression of non-linear control law (29) is quite large. In order to investigate the actual contribution of the reference path curvature $c(s)$ in the satisfactory guidance performances displayed on Figure 19, sine curve following has also been experimented with the simplified control law (30) (a priori dedicated to straight line following) where $c(s)$ has been set equal to 0. Reference path, tractor trajectory obtained with that law, and trajectory obtained with the entire control law (29), are displayed on Figure 20.

It can be observed that, without the reference path curvature information, the tractor follows the reference path with a 1.5 meters offset, and moreover presents large excursions when the values of $c(s)$ are the highest. This clearly establishes the relevancy of the reference path curvature information $c(s)$ used in non-linear guidance law (29).

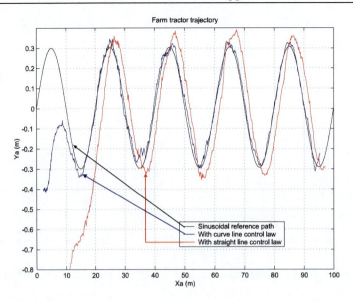

Figure 20. Tractor trajectories during a sine curve following: reference trajectory and trajectories obtained from control laws (29) and (30).

5.2. Guidance Laws Capabilities with Respect to Sliding Effects

Experiments reported in this Section have no longer been performed on asphalt but, as it is expected in agricultural tasks, on land fields.

When the vehicles move on a level compact ground, guidance results are absolutely similar to those presented in Section 5.1., whatever the guidance law is: since adaptive control law (35) and predictive law (46) have both been designed from non-linear control law (29), the performances and the attractive features of the latter one are propagated to the two former ones.

Therefore, experiments reported in this Section address specifically situations where sliding occurs. Vehicle behavior, and especially a loss in accuracy when automatic guidance relies on non-linear control law (29), is displayed. The relevancy of the approaches proposed to account for sliding effects in vehicle modeling and control (Model (11) and adaptive control law (35)), and to account for the actuator features (predictive control law (46)), are then demonstrated.

More precisely, three reference trajectories have been successively considered. The first one consists in a long curve to be achieved on a level slippery ground. Sliding effects occur when the vehicle enters into the curve, and since the reference path curvature is almost constant, the capabilities of the proposed control approaches with respect to quite constant sliding conditions can be investigated. The second reference path consists in a succession of half-turns on the same level slippery ground. Guidance accuracy, when sliding conditions are continuously varying, can then be addressed. The last reference path consists in a straight line to be achieved on a sloping field. Control laws capabilities, when sliding effects originate no longer from path curvature but from slope, can then be investigated. Additional experimental results can also be found in [8] and [9].

5.2.1. Guidance on Level Slippery Fields when the Path Curvature is Constant

Automatic guidance has first been achieved with respect to the reference path shown on Figure 21. This path, named below path #1, has not been computed from a GPS map of the field: the farm tractor has been manually driven, and the path has then been recorded from the position measurements provided by the RTK GPS sensor on-boarded. This is a simple and convenient method to be sure that the farm tractor can actually follow the assigned reference path.

This reference path consists mainly in a long curve (three quarters of a circle) achieved on a level slippery ground. Poor adherence conditions at wheels-ground contact points, reference path large curvature ($c(s) \approx 0.2\ m^{-1}$), and vehicle large inertia explain that sliding occurs when the farm tractor is describing the curved part of the path.

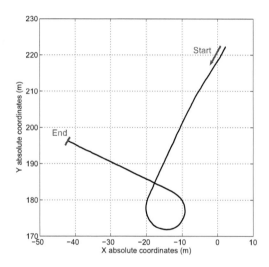

Figure 21. Reference path # 1.

Experimental validation of Model (11) and sliding estimation (16)

The capability of the vehicle extended kinematic Model (11) designed in Section 3.3. to account properly for sliding effects is first investigated via the experiments reported on Figure 22.

More precisely, path #1 following has been achieved at a constant velocity $v = 8\ km.h^{-1}$, relying on non-linear control law (29). Since the curvature is quite constant in the curved part of path #1, sliding effects are also expected to be constant. In that case, extended kinematic Model (11) predicts that lateral and angular deviations converge to some constant (see Section 4.2.), whose values could be computed from (31) and (34) if $\dot{Y}_P$ and $\dot{\Theta}_P$ values were known.

Lateral and angular deviations recorded during this experiment are depicted in solid line on Figures 22. The lateral deviation shown on Figure 22(a) fits exactly with the behavior predicted by Model (11): when the vehicle is describing the first straight line part of path #1 ($s < 45\ m$), its lateral deviation is close to zero, as it is expected from control law (29)

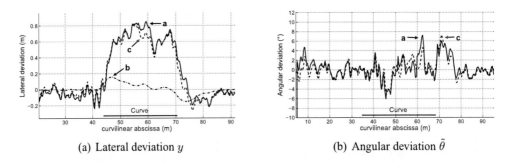

(a) Lateral deviation y (b) Angular deviation $\tilde{\theta}$

Figure 22. Deviations during path #1 following with control law (29): actual deviation (solid - *a*) and simulated ones from Models (10) (dash-dotted - *b*) and (11) (dashed - *c*).

since no sliding occurs. In contrast, when the vehicle enters into the curve ($s = 45\ m$), it undergoes sliding effects. It can be observed that a lateral deviation then appears, and remains quite constant ($y \approx 70\ cm$) as long as the vehicle is describing the curve. Finally, this lateral deviation returns to zero when the vehicle exits from the curve ($s > 72\ m$). The asymptotic value of the lateral deviation during the curved part of path #1 cannot however be verified, since we have no access to $\dot{Y}_P$ and $\dot{\Theta}_P$ values.

Figure 22(b) displays the angular deviation recorded during the same experiment. Observations are more difficult since the noise ratio on that measurement is more important than on lateral deviation measurement. This is always a consequence of the difficulty to have access to vehicle heading θ_t when relying solely on an RTK GPS sensor, as it has been extensively discussed above. Nevertheless, it can be noticed that during the straight line parts of path #1, the mean value of the angular deviation is zero, as it is expected from control law (29) since no sliding occurs. In contrast, when the vehicle is describing the curved part of path #1, the angular deviation then exhibits a non-null mean value, slightly inferior to 2°. This also agrees with what could be predicted from Model (11), see relation (31). The non-null mean value of angular deviation $\tilde{\theta}$ reveals that the vehicle is slightly moving crabwise.

The experiment reported on Figure 22 demonstrates that vehicle extended kinematic Model (11) can properly account for sliding effects, at least when these latter are constant. In order now to investigate if sliding estimation algorithm (16) can provide with satisfactory information, two simulations have been run:

- the first one, depicted in dash-dotted line on Figure 22, is path #1 following simulation, relying on Model (10), i.e. vehicle model without sliding accounted.

- the second one, depicted in dashed line on Figure 22, is also path #1 following simulation, but relying this time on Model (11). The perturbation information $\dot{Y}_P$ and $\dot{\Theta}_P$ introduced in that simulation have been provided by sliding estimation algorithm (16) that had been run during the above mentioned experiment, i.e. during path #1 following carried out with the farm tractor.

It can be observed on Figure 22(a) that the lateral deviation simulated from Model (11) fits almost perfectly with the actual lateral deviation recorded during the experiment:

- first, this simulation provides with the same 70 cm lateral deviation when the vehicle is describing the curved part of path #1,

- moreover, even the small variations of the actual lateral deviation (variations centered on 0 during straight line following, and centered on 70 cm when the farm tractor is describing the curve) are also reproduced on the simulation.

The angular deviation simulated from Model (11), shown on Figure 22(b), presents also satisfactory results:

- during straight line following, the simulated angular deviation also fits perfectly with the actual one: once more, even the small variations of the actual angular deviation are reproduced on the simulation.

- in contrast, when the vehicle is describing the curved part of path #1, the two curves are no longer overlapping. These differences originate from the method followed to measure the vehicle heading: raw vehicle heading measurement (13), as well as Kalman reconstructor equations (15), rely on pure rolling and non-sliding conditions at wheels-ground contact points. However, these conditions are no longer satisfied when the vehicle is describing the curved part of path #1. Therefore, in that situation, the vehicle heading reconstruction θ_t is not very accurate, which leads to the small inconsistencies noticed on Figure 22(b). Nevertheless, the simulated angular deviation presents the expected main feature, i.e. a non-null mean value.

In contrast, the simulation run from Model (10) provides with deviations that may be far from the actual ones. It can be observed on Figure 22(a) that:

- the lateral deviation simulated from Model (10) is exactly zero during straight line following. This is quite normal, since it is a purely theoretical simulation: no actual information is ever accounted in Model (10).

- this simulated lateral deviation presents only some overshoots when the vehicle is describing the curved part of path #1: these overshoots are generated by the delay introduced by the steering actuator (since actuator features are accounted in both simulations). They occur each time path #1 curvature presents fast variations (especially at the beginning/end of the curved part of path #1).

Since the lateral deviation simulated from Model (10) is close to zero even during the curved part of path #1, control law (29) (which has been designed from Model (10)) cannot be aware that special steering corrections are demanded in that case. This explains that in actual experiments, that control law leads to a non-satisfactory permanent lateral deviation. In contrast, the second simulation has revealed that vehicle extended kinematic Model (11), together with sliding estimation algorithm (16), can describe very accurately the vehicle behavior in presence of sliding. Therefore, this model and this estimation algorithm appear actually reliable to support control design refinements which could account for sliding effects, and then ensure a null lateral deviation in the curved part of path #1. These control designs have been presented in Sections 4.2. and 4.3.. Capabilities of the proposed guidance laws (35) and (46) are now investigated, still with respect to path #1 following experiments.

Guidance capabilities of adaptive control law (35)

First, path #1 following has been achieved, at constant velocity $v = 8\ km.h^{-1}$, relying on adaptive control law (35). The two approaches proposed in Section 4.2. in order to compute the adaptive corrective term y_c, i.e. MRAC and IMC schemes, have both been experimented. Guidance capabilities are displayed on Figure 23, and compared with those previously obtained with non-linear control law (29). More precisely, the conventions used on Figure 23 are:

- solid line: path following achieved from non-linear control law (29),

- dash-dotted line: path following achieved from adaptive control law (35), when the corrective term y_c is computed from MRAC scheme,

- dashed line: path following achieved from adaptive control law (35), when the corrective term y_c is computed from IMC scheme.

It can be observed on Figure 23(a) that, as expected, both adaptive control laws succeed in bringing back the vehicle lateral deviation close to zero during the curved part of path #1, despite sliding occurrence: when non-linear guidance law (29) leaves the vehicle $70\ cm$ far from its reference trajectory during the whole curved part of the path, both adaptive laws (35) provide a more satisfactory $15\ cm$ guidance accuracy during the main part of the same curve. More precisely, adaptive guidance law relying on IMC scheme ensures that, when $s \in [50\ m,\ 65\ m]$, the vehicle lateral deviation displays a satisfactory $6\ cm$ mean value and a $6\ cm$ standard deviation from the mean.

Nevertheless, it has also to be noticed on Figure 23(a) that adaptive control laws (35) cannot prevent from large transient guidance errors at the beginning/end of the curved part of path #1. At those places, path #1 curvature, and therefore sliding conditions, present fast variations. Due to the delay introduced mainly by the actuator device, the vehicle steering angle cannot react immediately, so that the vehicle transiently deviates from path #1. This very disappointing behavior has been addressed in Section 4.3., relying on Model Predictive control techniques. Capabilities of the proposed control law (46) with respect to these transient guidance errors are investigated in forthcoming experiments reports (see Figure 24).

It can also be observed on Figure 23 that the two adaptive schemes, i.e. MRAC and IMC, present very similar performances. From a theoretical point of view, it was expected that IMC scheme could enable to fit more accurately with the current sliding conditions, and therefore could provide with more accurate guidance capabilities: with IMC scheme, the adaptive corrective term y_c is always the most suitable one for any given perturbation values $\dot{Y}_P$ and $\dot{\Theta}_P$, since it is computed directly from the asymptotic relation (34). In contrast, with the MRAC scheme, perturbation values $\dot{Y}_P$ and $\dot{\Theta}_P$ are proceeded through a simulation application, which obviously acts as a low-pass filter when delivering y_c value.

This expected behavior for y_c values is actually obtained: Figure 23(d) displays y_c values provided by MRAC scheme in dash-dotted line as usual, and those directly provided by IMC scheme in solid line. It can be observed that, as soon as the vehicle enters into the curve, the y_c values supplied by IMC scheme are instantaneously close to the constant value

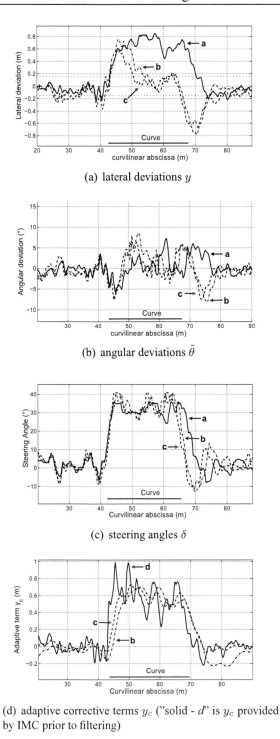

(a) lateral deviations y

(b) angular deviations $\tilde{\theta}$

(c) steering angles δ

(d) adaptive corrective terms y_c ("solid - d" is y_c provided by IMC prior to filtering)

Figure 23. Capabilities of non-linear control law (29) (solid - a), adaptive control laws (35) relying on MRAC scheme (dash-dotted - b) and IMC scheme (dashed - c).

that they should have during the whole curve (since path #1 curvature, and therefore sliding conditions, are constant). However, this positive feature is counterbalanced by a very negative one: the y_c values supplied by IMC scheme present a very poor noise ratio, and would obviously lead to oscillations on steering angle δ, if they were sent directly to adaptive control law (35). As it has been mentioned in Section 3.4.1., numerous perturbations, and not only sliding effects, are accounted in $\dot{Y}_P$ and $\dot{\Theta}_P$ values provided by estimation algorithm (16). Therefore, $\dot{Y}_P$ and $\dot{\Theta}_P$ present a poor noise ratio, and the same feature is propagated to y_c via relation (34). Consequently, as it has been announced in Section 4.2., the y_c values supplied directly from IMC scheme have to be filtered. The values actually used in adaptive control law (35) are shown in dashed line on Figure 23(d), and are then no longer very different from those provided by MRAC scheme. Therefore, guidance capabilities offered by the two schemes are quite similar, as it can be observed on Figure 23(a).

Finally, the vehicle behavior during curved part of path #1 can be further analyzed from Figures 23(b) and 23(c). When the vehicle is guided according to non-linear control law (29), its angular deviation does not present noticeable changes when the curve is described: it can just be noticed on Figure 23(b) that the angular deviation mean value becomes slightly non-null ($\approx 2°$) during the curve, indicating that the vehicle presents a slight crabwise motion. In contrast, when the vehicle is guided according to adaptive control law (35), its angular deviation presents higher values ($\approx 8°$) at the beginning of the curve, indicating that the vehicle is turning to join back reference path #1. This is also corroborated by higher steering angle values, as shown on Figure 23(c): adaptive control law (35) is clearly acting in order to reject sliding effects. Similar behavior can be observed when the vehicle is exiting from the curve. When it is guided according to non-linear control law (29), the vehicle is 70 cm aside from path #1 when the curve is described. When it reaches the end of this curve, no noticeable change is observed on its angular deviation $\tilde{\theta}$: the vehicle joins back path #1 just because sliding effects have disappeared. In contrast, when the vehicle is guided according to adaptive control law (35), it is close to path #1 when the curve is described. When it reaches the end of this curve, sliding effects suddenly vanish. Due to the delay introduced by the steering actuator, the vehicle transiently diverges from path #1, leading to a negative angular deviation, as it can be noticed on Figure 23(b), and negative steering angle values in order to bring back the vehicle on path #1, as it can be observed on Figure 23(c).

Guidance capabilities of predictive control law (46)

Finally, path #1 following has also been achieved, still at constant velocity $v = 8\ km.h^{-1}$, relying this time on predictive control law (46). The two possibilities for computing the adaptive corrective term y_c, i.e. MRAC and IMC schemes, have still been experimented. Guidance capabilities are reported on Figure 24, and are always compared with those previously obtained with non-linear control law (29). The convention used on Figure 24 are similar to the conventions previously introduced on Figure 23:

- solid line: path following achieved from non-linear control law (29),

- dash-dotted line: path following achieved from predictive control law (46), when the corrective term y_c is computed from MRAC scheme,

- dashed line: path following achieved from predictive control law (46), when the corrective term y_c is computed from IMC scheme.

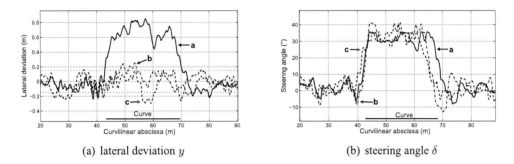

(a) lateral deviation y (b) steering angle δ

Figure 24. Capabilities of non-linear control law (29) (solid - a), predictive control laws (46) relying on MRAC scheme (dash-dotted - b) and IMC scheme (dashed - c)

Figure 24 demonstrates that, when the vehicles are guided according to predictive control laws (46), path #1 following can then be achieved without any large transient guidance error at the beginning/end of the curve. As detailed in Section 4.3., prediction only relies on reference path curvature information: the steering action is anticipated (with respect to the delay introduced by the actuator) when path #1 curvature is about to change, i.e. at the beginning/end of the curved part. Therefore, when the vehicle actually enters into the curve, or exits from the curve, actual steering angle δ_a displays a value consistent with path #1 current curvature. The adaptive corrective parameter y_c, still present in predictive control law (46), can then efficiently account for sliding effects as soon as they appear/disappear, so that an accurate path following can be achieved for the path long. Figure 24(a) shows that the vehicle lateral deviation belongs to the acceptance range of $\pm 15\ cm$ for most of the path #1 following experiment. The disturbance observed when $s \in [55\ cm, 60\ cm]$ is due to a hole on the ground crossed by the tractor, which is of course an unpredictable event.

Capabilities of the proposed guidance laws when the vehicles undergo sliding effects are summarized in Table 3. For all the guidance laws (excepted of course for non-linear control law (29)), these statistical computations have been carried out when the adaptive corrective term y_c is provided by IMC scheme (similar performances are however also obtained with MRAC scheme, see Figure 24(a)). More precisely:

- for non-linear control law (29) and predictive control law (46), the mean lateral deviation value and its standard deviation from the mean have both been computed when the vehicle is describing the curved part of path #1, i.e. when $s \in [45\ cm, 65\ cm]$.

- for adaptive control law (35), these two quantities have been computed only when the vehicle has joined back path #1 in the curved part of this path, i.e. when $s \in [52\ cm, 65\ cm]$. By this way, the statistical quantities are not polluted by actuator features and reflect actually guidance law capabilities with respect to sliding effects.

- the two other quantities in Table 3 provide with the range of variation of the vehicle lateral deviation. In the case of adaptive control law (35), they indicate the amplitude of the transient guidance errors due to steering actuator features.

Table 3. Statistical data on vehicle lateral deviation on the curved part of path #1

	mean value	std value	min value	max value
non-linear control law (29)	70 *cm*	9 *cm*	44 *cm*	85 *cm*
adaptive control law (35)	6 *cm* *(overshoots excepted)*	6 *cm*	-78 *cm*	59 *cm*
predictive control law (46)	-3 *cm*	12 *cm*	-30 *cm*	15 *cm*

Finally, Figure 24(b) reports steering angle values during path #1 following. The anticipation in the steering action, introduced by predictive control law (46), appears clearly when considering the experimental data recorded when the adaptive corrective term y_c is provided according to IMC scheme: steering angle δ is actually increasing prior to the beginning of the curve. Steering values are, as expected, anticipated with respect to the ones recorded when the vehicle is guided according to non-linear control law (29). The same steering behavior is observed at the end of the curve.

When the adaptive corrective term y_c is provided according to MRAC scheme, anticipation is not obvious at the beginning of the curve, just because the steering angle was accidently negative ($\approx -10°$) at the instant when the anticipation should have started. Moreover, since MRAC scheme accounts for sliding in a slightly slower fashion than IMC scheme, the steering angle must finally climb to higher values than with IMC scheme ($41°$ vs $35°$) in order to keep the vehicle on path #1. At the end of the curve, the anticipation provided by MRAC scheme can be observed, but higher negative steering values than with IMC scheme are however noticed, due once more to the way sliding is accounted.

5.2.2. Guidance on Level Slippery Fields when the Path Curvature is Varying

Path #1 following experiments have demonstrated that predictive control law (46) can ensure an accurate guidance when the vehicles undergo a constant sliding (sliding effects are expected to be constant, since the curved part of path #1 presents a constant curvature). In order to investigate guidance law (46) capabilities with respect to reference trajectories where sliding effects are no longer constant, automatic guidance has been achieved with respect to the path depicted on Figure 25, named below path #2.

This path has also been recorded from the position measurements provided by the RTK GPS sensor on-boarded, when the farm tractor has been once manually driven. Experiments have been carried out on the same slippery ground than for path #1. Therefore, the same sliding effects than those observed during path #1 following are qualitatively expected when the farm tractor is describing the curved parts of path #2. However, since path #2 presents both positive and negative curvatures, sliding effects change their direction during the completion of this path. Moreover, since the straight lines linking the half-turns are very short,

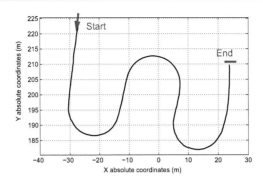

Figure 25. Reference path # 2.

the variations from positive sliding effects to negative ones are very fast. It is a very unfavorable situation that actually allows to investigate guidance laws capabilities when sliding conditions are varying.

Path #2 following has also been achieved at a constant velocity $v = 8 \ km.h^{-1}$. Adaptive control law (35) and predictive control law (46) have both been experimented. In both cases, the adaptive corrective term y_c has been supplied from IMC scheme. Non-linear control law (29) has also been experimented, in order to provide with a comparison. Vehicle lateral deviations are reported on Figure 26, with the following conventions:

- solid line: path following achieved from non-linear control law (29),

- dash-dotted line: path following achieved from adaptive control law (35), when the corrective term y_c is computed from IMC scheme,

- dashed line: path following achieved from predictive control law (46), when the corrective term y_c is computed from IMC scheme.

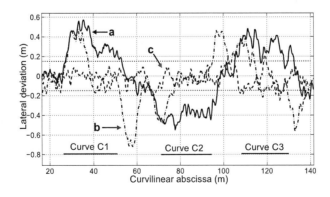

Figure 26. Lateral deviations recorded during path #2 following achieved from non-linear law (29) (solid - a), adaptive law (35) (dash-dotted - b), predictive law (46) (dashed - c).

First, it can be observed on Figure 26 that the vehicles guided according to non-linear control law (29) present large lateral deviations in each curved part of path #2, due obviously to sliding occurrences. The lateral deviation is successively positive and negative, depending on sliding effects direction. In contrast, guidance accuracy obtained with predictive control law (46) belongs satisfactorily to the acceptance range of $\pm 15\ cm$ during the whole path #2. Although the adaptive corrective term y_c used to account for sliding effects has been designed relying on a constant sliding assumption (see Section 4.2.), the online update of this parameter allows nevertheless to address satisfactorily situations where sliding effects are varying. Therefore, this proposed guidance law can actually deal with this very unfavorable situation, rarely encountered in agricultural tasks.

Figure 26 points also out the importance of the predictive part in control law (46). When this control refinement is not used, i.e. when automatic guidance is achieved according to adaptive control law (35), guidance accuracy is actually very poor: when the vehicle enters into the first curve, a large transient guidance error occurs, as it is expected since the steering actuator introduces a delay. Then, adaptive control law (35) accounts for sliding effects and brings the vehicle back on path #2. However, this path is constituted from a succession of short curves. Therefore, when the vehicle joins back the reference path, the first curve is already ended, and a second guidance overshoot occurs, and so on with the two other curved parts of path #2. Finally, due to the numerous overshoots, the guidance accuracy is almost as poor as when non-linear control law (29) is used. This demonstrates that, in order to follow with a satisfactory accuracy a path whose curvature is varying, it is mandatory to account for actuator features. Model Predictive Control techniques, used to design control law (46), have been shown to be a very efficient approach to address this problem.

5.2.3. Straight Line Following on a Sloping Field

In the above experiments, the farm tractor undergoes sliding effects since it is describing paths with large curvatures on a level slippery ground. Experiments reports have revealed that vehicle extended kinematic Model (11) and sliding estimation algorithm (16) can properly account for such sliding effects, and that the proposed control laws can provide with a satisfactory guidance accuracy, despite sliding occurrence.

However, vehicles can also undergo sliding effects when describing straight lines on a sloping field. The experiments reported below aim at demonstrating that the proposed modeling, sliding estimation and guidance law design are also relevant to address sliding effects originating from slope.

The last automatic guidance experiments have therefore been achieved with respect to the path shown on Figure 27, and named below path #3. This path is almost a straight line, achieved on a sloping field, perpendicularly to the slope (indicated by the arrows on Figure 27). All along this path, the field slope is slightly varying around a mean value equal to 15%. Poor adherence conditions at wheels-ground contact points, and gravity forces originating from vehicle large inertia and field slope, explain that sliding occurs permanently when the farm tractor is describing this path.

Experimental validation of Model (11) and sliding estimation (16)

In order to investigate the ability of vehicle extended kinematic Model (11) to account properly for sliding effects originating from slope, path #3 following has been achieved at a constant velocity $v = 6\ km.h^{-1}$, relying on non-linear control law (29). Since the

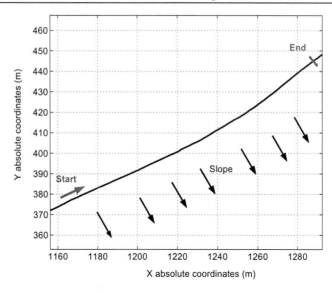

Figure 27. Reference path #3.

slope is quite constant all along this reference path, sliding effects are also expected to be constant. Therefore, it is finally the same situation than the one obtained when the vehicles are describing a path with a constant curvature on a level slippery ground, although sliding effects are not originating exactly from the same physical phenomena. As it has been mentioned above, extended kinematic Model (11) predicts in that case that the vehicle lateral and angular deviations converge to some constants (see Section 4.2.), whose values could be computed from (31) and (34) if $\dot{Y}_P$ and $\dot{\Theta}_P$ values were known.

The deviations recorded during this experiment are depicted in solid line on Figures 28. The vehicle lateral deviation shown on Figure 28(a) does not present a constant non-null value, as it was expected: it can be observed that the vehicle is oscillating within a very large 80 cm strip aside from path #3. It is however immediate to note that this strip is definitely not centered on the reference path. Therefore, if only the mean value of the lateral deviation is regarded, a large non-null value is then obtained (close to 30 cm), which is somehow consistent with what was expected from Model (11).

The large oscillations displayed by the vehicle can be explained by:

- the field slope, which is probably not perfectly constant along path #3,
- the tractor cabin oscillations: when a vehicle is moving perpendicularly to a slope, the mass distribution on the vehicle is noticeably modified. Therefore, the two shock absorbers located on the opposite side from the slope are more compressed than the two others, thus leading to tractor cabin oscillations. Since the GPS antenna is located on the top of the cabin, position measurements provided by the RTK GPS sensor are likely to present also some oscillations, which are directly reflected on lateral deviation measurements shown on Figure 28(a).

The vehicle angular deviation recorded during the same experiment is displayed on Figure 28(b). It has been abundantly mentioned above that vehicle heading measurement

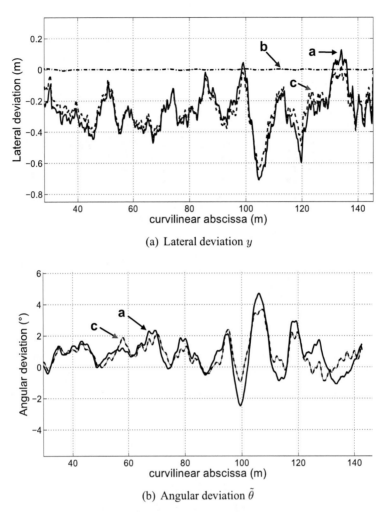

Figure 28. Deviations during path #3 following with control law (29): actual deviation (solid - *a*) and simulated ones from Models (10) (dash-dotted - *b*) and (11) (dashed - *c*).

is a critical point in our application, even in the most favorable situation, since it relies on a single RTK-GPS sensor. With respect to vehicle heading measurement, the current experiment is undoubtedly far from the most favorable situation: the tractor cabin presents oscillations, pure rolling and non-sliding conditions at wheels-ground contact points (which are explicitly used to derive θ_t) are not satisfied, and moreover sliding effects are varying since the field slope is probably not constant. Therefore, it is not surprising to observe a very oscillating angular deviation $\tilde{\theta}$ on Figure 28(b). However, if once more only the mean value of the deviation is regarded, a non-null value is clearly obtained ($\approx 1°$), which is again consistent with what could be predicted from Model (11), see relation (31). This non-null mean value for the angular deviation $\tilde{\theta}$ reveals that, as in path #1 following, the vehicle is slightly moving crabwise.

Vehicle extended kinematic Model (11) appears therefore consistent with the experimental results reported on Figure 28, at least if only mean deviation values are regarded.

In order now to investigate if sliding estimation algorithm (16) can provide with satisfactory information, two simulations have again been run. They are exactly similar to those achieved when investigating sliding estimation algorithm (16) with respect to path #1. The conventions used on Figure 28 are also similar:

- path #3 following simulation relying on Model (10) (i.e. vehicle model without sliding accounted) is shown in dash-dotted line.

- path #3 following simulation relying on Model (11) is shown in dashed line. The perturbation information $\dot{Y}_P$ and $\dot{\Theta}_P$ introduced in that simulation are provided by sliding estimation algorithm (16) that had been run during the above mentioned experiment, i.e. when the farm tractor was achieving path #3 following.

As it was expected, Model (10) cannot describe the actual vehicle behavior: since no actual information is accounted in Model (10), this simulation is a purely theoretical one, providing with a constantly null simulated lateral deviation, when the actual lateral deviation is non-null and varying, see Figure 28(a).

In contrast, it can be observed on Figure 28(a) that the lateral deviation simulated from Model (11) fits perfectly with the actual lateral deviation recorded during the experiment: even the small variations of the actual lateral deviation are reproduced on the simulation. Simulated and actual angular deviations are also satisfactorily consistent, as it is displayed on Figure 28(b). Small differences can be noticed, but they are inherent to the vehicle heading measurement method, which is not completely correct when the vehicles undergo sliding effects (this point has already been discussed when sliding estimation algorithm (16) had been investigated with respect to path #1). Therefore, extended kinematic Model (11), together with sliding estimation algorithm (16), appear actually reliable to describe the vehicle behavior, not only when sliding effects originate from reference path curvature, but also when they originate from slope.

Adaptive and predictive guidance laws (35)-(46), designed from Model (11), have been shown, in the above reported experiments, to be efficient when sliding effects originate from reference path curvature. It is shown below that they can also provide with satisfactory guidance results when sliding effects originate from slope.

Guidance capabilities of adaptive control laws (35)

Path #3 following has therefore been achieved, at constant velocity $v = 6\ km.h^{-1}$. Predictive control laws (46), designed in Section 4.3., aim at providing anticipation to the steering action from reference path curvature information. However, since path #3 consists in a straight line, no anticipation can actually be provided, so that predictive control laws (46) are exactly identical to adaptive control laws (35) in this application. Therefore, only experimental results obtained with these latter guidance laws are shown on Figure 29. Once more, the two approaches proposed in Section 4.2. in order to compute the adaptive corrective term y_c, i.e. MRAC and IMC schemes, have both been experimented. Guidance capabilities obtained with non-linear control law (29) are also reported in order to provide with a comparison. The conventions used on Figure 29 are similar to the conventions previously used on Figure 23 to display path #1 following reports:

- solid line: path following achieved from non-linear control law (29),

- dash-dotted line: path following achieved from adaptive control law (35), when the corrective term y_c is computed from MRAC scheme,

- dashed line: path following achieved from adaptive control law (35), when the corrective term y_c is computed from IMC scheme.

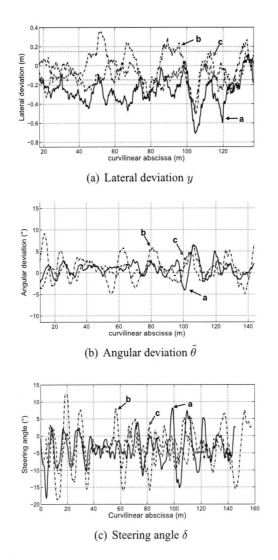

(a) Lateral deviation y

(b) Angular deviation $\tilde{\theta}$

(c) Steering angle δ

Figure 29. Capabilities of non-linear control law (29) (solid - *a*), adaptive control laws (35) relying on MRAC scheme (dash-dotted - *b*) and IMC scheme (dashed - *c*).

It can be observed on Figure 29(a) that, whatever the guidance law is, the farm tractor describes the reference straight line with noticeable oscillations. The measurement device on-boarded on the tractor has a large responsibility in this oscillating behavior: in such an unfavorable situation, this device relying solely on an RTK-GPS sensor, is not able to provide with accurate enough state measurements:

- the GPS antenna is located on the top of the tractor cabin. Therefore, due to the above mentioned tractor cabin oscillations, the position measurements supplied by the RTK-GPS sensor are obviously not as reliable as they were in guidance applications achieved on a level ground.

- raw vehicle heading measurements θ_t are, according to (13), computed from two successive position measurements. Therefore, the slight inaccuracy in these latter measurements is also propagated to raw θ_t measurements.

- finally, both raw vehicle heading measurements (13) and Kalman reconstructor equations (15) have been derived according to pure rolling and non-sliding conditions at wheels-ground contact points. However, when a vehicle is moving on a sloping field, these conditions are constantly unsatisfied. This is corroborated by Figure 29(b): the vehicle angular deviation is continuously presenting a non-null value, indicating that the vehicle is continuously moving crabwise (this occurs only from time to time when the vehicle was describing a curve on a level ground).

This less reliability of position and heading measurements can explain for a large part the oscillating feature exhibited by the vehicle lateral deviation on Figure 29(a). Nevertheless, it is worth noticing that both adaptive control laws (35) provide, when compared with non-linear control law (29), with a significatively more accurate vehicle guidance. To quantify this point, the time spent by the farm tractor within a $\pm 15\ cm$ strip centered on the reference path has been computed for each guidance law. These data have then been converted into a percentage and reported in Table 4, which displays clearly the improvement in guidance accuracy.

Table 4. Guidance accuracy during path #3 following

	time spent within a $\pm 15\ cm$ strip centered on path #3
with non-linear control law (29)	16 %
with adaptive control law (35) and MRAC scheme	59 %
with adaptive control law (35) and IMC scheme	70 %

It can be observed on Table 4 that adaptive control law (35) relying on IMC scheme leads to a more accurate path #3 following than when it relies on MRAC scheme (70 % vs. 59 %). The difference in the performances of these two guidance laws is even more significant when considering Figure 29(a): when adaptive guidance law (35) relies on MRAC scheme, the farm tractor presents larger oscillations than when it relies on IMC scheme. These larger oscillations originate from the delay in accounting properly for sliding effects, that it is slightly larger with MRAC scheme than with IMC scheme: in MRAC scheme, the adaptive corrective term y_c is obtained from a simulation application. Changes in sliding parameters $\dot{Y}_P$ and $\dot{\Theta}_P$ require then some delay prior to be reflected in y_c value. In contrast, in IMC scheme, the adaptive corrective term y_c is derived from asymptotic relation (34).

Therefore, even if this y_c value has to be filtered, changes in sliding parameters $\dot{Y}_P$ and $\dot{\Theta}_P$ can be reflected within a shorter time. When the vehicles are moving on a sloping field, sliding conditions change in a faster way than when they are describing curves on a level ground. Therefore, differences between IMC and MRAC schemes, that were not obvious in the latter situation, appear more clearly in the former one. The small delay introduced by MRAC scheme when accounting for sliding, is compensated via higher steering angle values, as it can be observed on Figure 29(c). Higher steering angle values lead necessarily to higher angular deviations, see Figure 29(b). Therefore, the vehicle trajectory presents more oscillations when MRAC scheme is used, rather than when adaptive control law (35) relies on IMC scheme, as it can be observed on Figure 29(a).

6. Conclusion

The aim of this work was to provide agricultural vehicles with automatic guidance capabilities, relying on an RTK GPS as the single vehicle sensing device. More precisely, accurate curved path following on agricultural fields where sliding is likely to occur, was addressed.

Since sliding effects have to be taken into account, a natural approach would have been to design guidance laws from a vehicle dynamic model. However, in as harsh conditions as are agricultural ones, these very large models, enclosing numerous badly known and even varying parameters, do not appear very tractable. It has been shown here that, as an alternative, sliding effects on vehicle motion can be satisfactorily accounted by an extended kinematic model, derived by introducing an additive structured perturbation into a vehicle kinematic model designed under pure rolling and non-sliding conditions at wheels-ground contact points.

Capability of the single RTK GPS sensor to provide online the state and perturbation vectors of that extended kinematic model has then been discussed. State vector, i.e. vehicle relative location and relative heading with respect to the reference path, can clearly be inferred from the position measurements supplied by this sensor, but heading information have nevertheless to be proceeded through a Kalman state reconstructor in order to improve its accuracy and then enable its use inside guidance laws. Perturbation vector can also be satisfactorily estimated, by comparing the actual vehicle position supplied by the RTK GPS sensor, with its expected position when sliding does not occur, obtained from online simulation.

Vehicle guidance laws design, relying on that extended kinematic model, has then been carried out according to three steps. First, it has been assumed that the vehicles do not slide. In that case, their kinematic model can be converted into a so-called chained-form, from which very attractive non-linear control laws can be designed: high accuracy curved path following can then be achieved, since both vehicle model non-linearities and path curvature can be explicitly taken into account. In addition, although these guidance laws are non-linear ones, performances tuning can still be achieved as intuitively as in the linear systems case. Moreover, these performances can be set independently from the vehicle velocity, which is a very interesting feature from a practical point of view.

Next, since in vehicle extended kinematic model, sliding effects are described as a perturbation term, it has been possible to account for them via adaptive control techniques. More precisely, relying on internal model approach, an adaptive corrective term has been introduced into previous non-linear control laws: guidance accuracy can then be ensured despite sliding occurrence, at least when sliding effects are slowly varying, while all the attractive features of non-linear guidance laws are still preserved.

Finally, in order to ensure guidance accuracy even when sliding effects present large variations (e.g. at the beginning/end of a curve), some anticipation has to be provided to the control laws, in order to counterbalance for the delays enclosed in the guidance feedback loop, mostly due to the actuator device. Future values of the reference path curvature are obvious data on which anticipation can efficiently rely. Model predictive control techniques have been shown to be a relevant approach to incorporate such an anticipation into the above designed guidance laws, and therefore to preserve their accuracy even at curvature path discontinuities.

Capabilities of the proposed guidance laws have been extensively investigated via numerous full-scale experiments. The first ones have been carried out on an asphalted ground. The aim was to achieve path following in a very favorable situation where the farm tractor does not undergo sliding, so that all the attractive features of the non-linear control laws can easily be highlighted. Then, experiments on agricultural fields have been performed, and much more unfavorable situations, where the farm tractor undergoes severe sliding, have been considered: namely, sharp curve following on a level slippery ground and straight line following on a sloping field. These experiments have demonstrated the capabilities of the adaptive control scheme to preserve guidance accuracy despite sliding occurrence, and abilities of predictive control scheme to avoid transient guidance error at sliding discontinuities.

Current continuation of this research work aims at introducing some additional improvements, on one hand by addressing modeling refinements, and on the other hand by extending at minimum vehicle sensing device.

Modeling refinements are first investigated in order to provide with more accurate values for the vehicle heading. For the present moment, both raw vehicle heading measurements and Kalman state reconstructor equations rely on pure rolling and non-sliding conditions at wheels-ground contact points, which is obviously not completely satisfactory. In some unfavorable situations, such as straight line following on a sloping field, the accuracy of vehicle heading values is even significantly damaged, since the vehicle is continuously moving crabwise, as it has been observed on the experimental reports. Another concern when measuring vehicle heading are tractor cabin oscillations: tractor roll and pitch are not accounted for the present moment. However, they slightly damage position measurements, and by propagation, damage more seriously vehicle heading values, especially when the vehicles are moving on irregular ground. Therefore, more comprehensive models, accounting for these effects, are currently considered in order to support vehicle heading evaluation, that would be derived from these models according to observers (such as Luenberger).

Modeling refinements are also addressed with the aim to obtain more representative values for sliding variables. For the present moment, they are derived from the comparison between the vehicle actual position, provided by the RTK GPS sensor, and the vehicle expected position if it had not undergone sliding, supplied from online simulation. Such an estimation is however not always totally relevant, since any perturbation on vehicle position measurement (and firstly when the tractor undergoes roll and/or pitch) is directly propagated to sliding variables values: for instance, when a vehicle crosses a hole on the ground, the tractor cabin is shaken, the vehicle actual and expected positions do not fit, and therefore sliding is detected when it has not to be. Other dynamic phenomena, such as delays originating from the vehicle large inertia, can also generate inconsistencies between the vehicle actual and expected positions, and then lead to misinterpreted sliding effects. Therefore, more comprehensive models which could provide with more representative values for sliding variables, are also currently investigated.

Finally, modeling refinements are also developed in order to improve the efficiency of prediction algorithms. At the present moment, only the steering actuator model is explicitly used when deriving the anticipation to be introduced in guidance laws, with respect to the future values of the reference path curvature. However, some other dynamic phenomena also generate delays in guidance feedback loop, and firstly those originating from the large vehicle inertia. In the experiments reported above, they are implicitly accounted since the prediction horizon has been empirically extended. Nevertheless, a more efficient anticipation could obviously be provided if more comprehensive models, incorporating these dynamic features, were used explicitly in prediction algorithms.

All the above mentioned modeling refinements are devoted to estimation and prediction purposes: they aim at accounting explicitly for dynamic features in order to either improve the accuracy of some variables describing the vehicle configuration, or to improve the models used in prediction algorithms. Vehicles dynamic models are however not expected to be used for control design purposes: the extended kinematic model presented here appears actually very tractable, since it allows to benefit from all the attractive features of guidance laws designed when it is assumed that the vehicles do no slide, while nevertheless accounting very efficiently for sliding effects.

In order to extract reliable information from these partial vehicle dynamic models, some additional measurements have clearly to be introduced: it does not appear possible to reconstruct with a satisfactory accuracy a larger vehicle attitude solely from the 3D position measurements supplied by a single RTK GPS. A minimum additional sensing equipment, which would enable to design efficient observers from partial dynamic models, is currently investigated. Theoretical developments (such as evaluation of observability rank conditions) and experimental characterization of the performances of some sensors in agricultural conditions, are being carried out.

References

[1] T. Bell, M. O'Connor, V.K. Jones, A. Rekow, G. Elkaim, and B. Parkinson. Realistic autofarming closed-loop tractor control over irregular paths using kinematic GPS. In *Proc. of the European Conference on Precision Agriculture (ECPA)*, 1997.

[2] P. Borne, G. Dauphin-Tanguy, F. Rotella, and I. Zambettakis. *Commande et optimisation de processus*. Series: Méthodes et pratiques de l'ingénieur. TECHNIP Editions, Paris, 1990.

[3] C. Debain, T. Chateau, M. Berducat, P. Martinet, and P. Bonton. An help guidance system for agricultural vehicles. *Computers and Electronics in Agriculture*, **25**(1-2):29–51, 2000.

[4] E. Dormegnie, G. Fandard, G. Mahajoub, and F. Zarka. *Dynamique du véhicule.* Lectures at French Institute for Advanced Mechanics (IFMA), Clermont-Ferrand, France, 2002.

[5] M. Ellouze and B. d'Andréa Novel. Control of unicycle-type robots in the presence of sliding effects with only absolute longitudinal and yaw velocities measurement. *European Journal of Control*, **6**:567–584, 2000.

[6] A. Gelb. *Applied optimal estimation.* MIT Press, Cambridge, MA, 1974.

[7] T. Holzhüter and R. Schultze. Operating experience with a high precision track controller for commercial ships. *Control Engineering Practice*, **4**(3):343–350, 1996.

[8] R. Lenain, B. Thuilot, C. Cariou and P. Martinet. Adaptive control for car-like vehicles guidance relying on RTK-GPS: rejection of sliding effects in agricultural applications. In *Proc. Intern. Conf. on Robotics and Automation (ICRA)*, pages 115–120, Taipei, 2003.

[9] R. Lenain, B. Thuilot, C. Cariou and P. Martinet. Rejection of sliding effects in car-like robot control. Application to farm vehicle guidance using a single RTK GPS sensor. In *Proc. Intern. Conf. on Intelligent Robots (IROS)*, pages 3811–3816, Las Vegas, 2003.

[10] H. Loeb. *Engin nautique de surface robotisé : synthèse de lois de commande et GPS différentiel.* PhD thesis, University of Bordeaux I, France, 1996.

[11] R. Lozano and D. Taoutaou. *Identification et commande adaptative.* Hermes, Paris, 2001.

[12] Y. Nagasaka, R. Otani, K. Shigeta, and K. Taniwaki. Automated operation in paddy fields with a fiber optic gyro sensor and GPS. In *Proc. International Workshop on Robotics and Automated Machinery for Bio-Production (Bio-Robotics)*, pages 21–26, Valencia (Spain), 1997.

[13] M. O'Connor, G. Elkaim, T. Bell, and B. Parkinson. Automatic steering of a farm vehicle using GPS. In *Proc. International Conference on precision agriculture*, pages 767–777, Mineapolis (USA), 1996.

[14] J. Richalet. Industrial applications of model based predictive control. *Automatica*, **29**:1251–1274, 1993.

[15] J. Richalet. *Pratique de la commande prédictive.* Traité des nouvelles technologies série automatique. Hermes, Paris, 1993.

[16] J. Ried and D. Niebuhr. Driverless tractors. *Ressource*, **8**(9):7–8, 2001.

[17] C. Samson. Control of chained systems. Application to path following and time-varying point stabilization of mobile robots. *IEEE Transactions on Automatic Control*, **40**(1):64–77, 1995.

[18] H. Sussmann, E. Sontag, and Y. Yang. A general result on the stabilization of linear systems using bounded controls. *IEEE Transactions on Automatic Control*, **39**(12):2411–2425, 1994.

[19] B. Thuilot. *Contribution à la modélisation et à la commande de robots mobiles à roues*. PhD thesis, Ecole des Mines de Paris, France, 1995.

[20] O. Yukumoto, Y. Matsuo, and N. Noguchi. Robotization of agricultural vehicles. *Japan Agricultural Research Quaterly (JARQ)*, **34**(2), 2000.

[21] The Zodiac. *Theory of robot control*. C. Canudas de Wit, B. Siciliano, G. Bastin editors, Springer-Verlag, Berlin, 1996.

In: Global Positioning Systems   ISBN 978-1-60741-012-6
Editors: V. Asphaug and E. Sørensen, pp. 117-156   © 2010 Nova Science Publishers, Inc.

Chapter 6

GLOBAL POSITIONING SYSTEM CONSTRAINTS ON PLATE KINEMATICS IN THE SOUTHERN ALPS AT THE NUBIA-EURASIA BOUNDARY

M. Bechtold[1], D. Zuliani[2], P. Fabris[2]
D.C. Tanner[3] and M. Battaglia[4]
[1]Institute of Chemistry and Dynamics of the Geosphere,
Institute Agrosphere, Research Centre Juelich, Germany
[2]Centro Ricerche Sismologiche,
Udine, Italy
[3]Leibniz Institute for Applied Geosciences,
Hannover, Germany
[4]Department of Earth Sciences, La Sapienza University,
Rome, Italy

Abstract

We present and interpret deformation velocities from continuous Global Positioning System (GPS) observations at 42 sites around the Adriatic region (central Mediterranean) to investigate the active tectonics of the Nubia-Eurasia plate boundary in the northeast of Italy, the seismically most active region of the Southern Alps. In total, 1600 days of GPS observations, from June 2002 to November 2006, were processed using GAMIT/GLOBK. In detail, we present our processing scheme that is based on a three-step approach: i) estimation of station coordinates, atmospheric zenith delay, orbital and Earth orientation parameters using daily GPS observations and applying loose constraints to geodetic parameters; ii) combining, on a daily basis, the loosely-constrained solutions with loosely-constrained SOPAC solutions; and combining the daily combinations into monthly averages while adjusting the final χ^2 to 1 by rescaling the daily h-files (method to account for white noise); and iii) defining the reference frame, determining the site velocities and estimating the error by including both the effect of white noise and random walk component (monument instability). We estimate the random walk component station by station using the "realistic sigma" algorithm of Herring (2003b).

All eight stations of the Friuli Regional Deformation Network (FReDNet), which form the focus of this chapter, present at least 2.5 years of data, and so exceed the minimum time span for a reliable determination of deformation velocity. The geodetic data indicate shortening of the crust in the region, with southern Friuli moving NNW towards northern Friuli at the relative speed of 1.6 to 2.2 mm/yr. One of the eight processed FReDNet stations (i.e. CANV - Caneva), situated closely north to the active Polcenigo-Maniago Thrust, is moving faster than its expected long-term, geologic velocity. As there are no other plausible reasons for this discrepancy, we interpret this anomalous high velocity to be an indication of a locked thrust and predict that strain is accumulating to the north of it.

1. Introduction

In the Italian Southern Alps, seismicity is mostly of low magnitude and quite scarce. An exception is the Eastern Southern Alps region, the site of frequent and strong seismic activity, as recently emphasized by the disastrous 1976 Friuli seismic sequence. In May and September 1976, the region was struck by two earthquakes of magnitude 6.5. It suffered 989 fatalities, 45 000 people became homeless and there were devastating economic losses [Slejko et al. 1999], all demonstrating the high vulnerability of the region due to high population density and density of industrial settlements. During the last 1 000 years the Eastern Southern Alps (NE Italy, Veneto and Friuli regions) were affected by at least eight earthquakes with magnitudes between 6 and 7 [Working Group CPTI 1999]. All of them had severe consequences for the region, causing damage from intensity I = 9 (ruinous) to I = 10 (disastrous) on the Mercalli-Cancani-Sieberg (MCS) scale. The reason for these ground motions is the independent movement of the Adriatic microplate within the Africa-Eurasia collision zone.

Since the Late Cretaceous (ca. 70 Ma), the Nubia and Eurasian plates have been converging, currently the rate of convergence is rather slow compared to geologic times [DeMets et al. 1994, Calais et al. 2003]. Roeder [1985] proposes a total shortening between Africa and Europe of about 400 km in the past 70 Ma. This would approximately correspond to an average velocity of ca. 6 mm/yr. After an early phase of subduction of oceanic lithosphere, i.e. the Alpine Jurassic Tethys Ocean, continental collision dominated from the Eocene onwards [Vannucci et al. 2004]. For Serpelloni et al. [2007], the Western-Central Mediterranean today provides a geodynamic example of the final stages of a continent-continent collisional orogeny. Due to weak GPS coverage in Africa and the tectonic complexity in the Mediterranean, which represents the westernmost portion of the Alpine-Himalayan orogenic belt, the angle of collision is the subject of ongoing controversial discussions (see Serpelloni et al. 2002, and references therein). However, geologic and geophysical studies mostly propose a northwestward motion of the Nubia plate (see Nocquet and Calais 2004). The NUVEL-1A plate motion model predicts N–S convergence at a velocity of 10 mm/yr in the western Mediterranean that turns gradually into a NW–SE convergence at lower velocities of ca. 4 mm/yr in the Gibraltar Strait [DeMets et al. 1994].

The long-lasting interaction of the Eurasian and Nubia (and also Arabian) plates produced a complex pattern of active deformation in the Mediterranean, with deep basins and looping fault-and-thrust belts, produced under a variety of stress states, from compressive to extensional, and associated strike-slip faulting [Dewey et al. 1973, 1989, Serpelloni et al. 2002]. The Mediterranean is now considered to be an assemblage of several lithospheric blocks that interact and move independently between the rigid Nubia and Eurasian plates. The fragments (e.g., Adria and Arabia) are mostly assumed to be former promontories derived from an irregularly-shaped, northern continental margin of the Nubia plate. In compressive stress regimes, when smaller plates collide, such as the case of the Alpine and Carpathian orogenic belts, a thickened continental lithosphere related to intense thrusting has developed. Zones of extensional deformation are characterized by a thinned continental lithosphere, as in the back-arc basins of the Aegean and Tyrrhenian subduction zones [Vannucci et al. 2004].

The Nubia-Eurasia plate boundary is characterized by a broad belt of seismicity. The complexity is caused by the previously described fragmentation into smaller plates and the weaker rheology of continental lithosphere, compared to the one of oceanic lithosphere. Due to its higher rigidity, oceanic lithosphere is less or weakly deformed and the deformation is mainly concentrated in the adjacent continental areas [Serpelloni et al. 2007].

1.1. The Adriatic Microplate

Within the Mediterranean, the Po Plain – Adriatic region forms a quasi-aseismic block that led to the suggestion of the presence of a smaller tectonic element, the so-called Adriatic microplate, or simply 'Adria' [Anderson and Jackson 1987]. Adria is built up of continental lithosphere and is surrounded by seismogenic zones on its western, northern and eastern sides, the circum-Adriatic seismic belt. N–S compression with thrust faulting characterizes the Eastern Southern Alps in the north, whereas the Dinarides and Albanides to the east of Adria demonstrate both thrust and strike-slip faulting. The western side is dominated by NE–SW extension along normal faults in the Apennines [Vannucci et al. 2004, Serpelloni et al. 2005]. To the south, the Adriatic microplate presents a significant lack of seismicity, leading to a poorly defined, diffuse southern boundary [Anderson and Jackson 1987, Battaglia et al. 2004].

Anderson and Jackson [1987] first proposed an anti-clockwise rotation of the Adriatic microplate with respect to the Eurasian plate. The anti-clockwise rotation was confirmed by Battaglia et al. [2004], who developed a block model of the Adriatic region to answer the fundamental question of whether Adria is a promontory of Nubia or an independent microplate. The model in which Nubia and Adria are not connected, describes the GPS velocities better than a model that considers these two regions as a single block. Proposed locations of the Euler pole of rotation vary between the Western Alps [Anderson and Jackson 1987] and the western part of the Po Plain [Calais et al. 2002, Battaglia et al. 2004]. The available continuous GPS deformation velocities (relative to a stable Eurasia) decrease across the Adriatic region from south (ca. 5 mm/yr) to north (ca. 2 mm/yr). The anti-clockwise rotation of the Adriatic microplate causes an increasing convergence from the western to the eastern Alps. The band of higher seismicity in the middle of the Adriatic microplate, originating from activity on the Gargano-Dubrovnik Fault (see Figure 1), was used by Oldow et al. [2002] to split the Adria into two blocks. These blocks correspond to the North Adria and South Adria in Battaglia et al. [2004; Figure 1], who tested this configuration and concluded that there was no geodetic evidence for the existence of two independent blocks. Argnani [2006] also contradicted a division in two blocks and points out that there is no geologic information to support a single major discontinuity that acts as a boundary at lithospheric scale. Furthermore, he questioned the general assumption of rigid behavior of the Adria and proposed significant intraplate deformation.

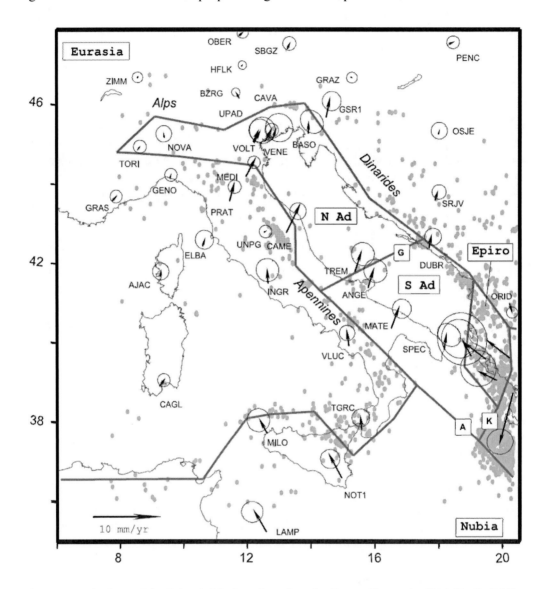

Figure 1. Block model of the Adriatic microplate by Battaglia et al. [2004]. Solid lines define the geometry of the modeled blocks. The arrows represent the horizontal GPS velocities and their 95 % confidence ellipses, referenced to the stable Eurasian frame realized by McClusky et al. [2000]. The grey dots indicate the location of the shallow seismicity from 1975 to 2000 (M > 3.5). Abbr.: N Ad: North Adria, S Ad: South Adria. G: Gargano-Dubrovnik Fault Zone; K: Kefallinia Fault Zone; A: Apulia Escarpment.

1.2. The Active Faults of Friuli

Results of the recently-performed seismic TRANSALP profile in the Eastern Alps are consistent with deep under-thrusting and wedge indentation of the Adriatic lithosphere under

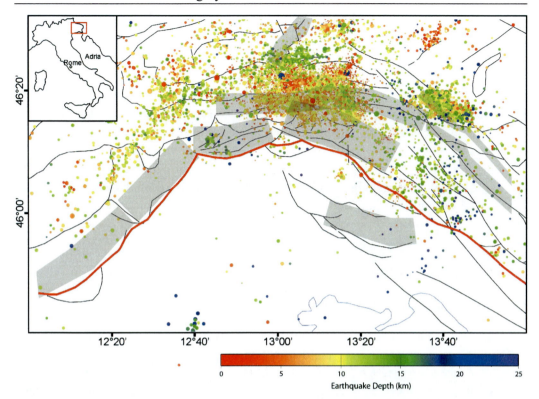

Figure 2. Seismicity distribution in Friuli. Gray polygons represent proposed seismogenic sources after DISS Working Group [2006]. Seismicity distribution is shown as dots scaled to magnitude and colored by depth. The structural map of Galadini et al. [2005] is also shown. The red line represents the trace of the modeled thrust fault. Inset shows the position of the study area in Italy [Bechtold et al. submitted].

the southern margin of the Eastern Alps thrust-and-fold belt [Castellarin et al. 2006]. The ongoing N-S to NW-SE compression [e.g., Bressan et al. 1998; D'Agostino et al. 2005], with faulting mainly along NE-SW to NW-SE-trending thrusts and the deformation of the Quaternary deposits of the Friulian Plain [Galadini et al. 2005] indicate that the Eastern Southern Alps can be presently described as a S-verging, foreland-propagating thrust system. According to D'Agostino et al. [2005], the motion of Adria is entirely absorbed in the southern Alps through thrusting and crustal thickening, with very little or no motion transferred to the north.

Knowledge of the related seismogenic sources is sparse [e.g., Bressan et al. 1998, 2003; Slejko et al. 1999; Poli et al. 2002; Poli and Renner 2004; Galadini et al. 2005; DISS Working Group 2006, see Figure 2]. During the last three decades, however, seismic recording has revealed new insights in the tectonics of the region. The recent tectonic activity is suspected to be related to motion on blind thrusts, some of them covered by the Quaternary deposits of the Friulian Plain. A major controlling factor is ascribed to the complex paleofault system of the area. Several Cenozoic compressional tectonic phases related to the Alpine Orogeny caused SW-NE-trending faults in the western, E-W-trending faults in the central, and SE-NW-trending faults in the eastern sector [Bressan et al. 1998, 2003; Galadini et al. 2005]. The seismicity distribution observed since 1977 cannot only be related to one active thrust front (Figure 2). Beside the proposed active faults, the area is probably characterized by other, as yet not recognized, active structures.

Since 6 years, high-precision geodesy, i.e. the Friuli Regional Deformation Network (FReDNet), monitors the movements of the crust in the Friuli area and provides an important tool for characterizing the ongoing deformation in Friuli and for identifying its active structures.

In the following section we give a detailed presentation of our processing of the GPS data of FReDNet. The resulting velocity field of Friuli is used to define the zone of shortening, the shortening rate, the angle of collision, and possible strain accumulation along faults.

2. Friuli GPS Velocity Field

2.1. The Continuous GPS Network FReDNet

The Friuli Regional Deformation Network (FReDNet) is a network of continuously operating GPS receivers monitoring the distribution of crustal velocities in the Friuli area [http://www.crs.inogs.it; see also Battaglia et al. 2003]. The geodetic network is operated by the 'Centro Ricerche Sismologiche' (CRS) of the 'Istituto Nazionale di Oceanografia e di Geofisica Sperimentale' (OGS). Network installation and data analysis began in summer 2002, and currently (September 2008) 12 permanent GPS stations (ACOM, AFAL, CANV, CODR, FUSE, JOAN, MDEA, MPRA, PAZO, TRIE, UDI1, ZOUF, see Figure 3) are up and running. In this chapter, we present results of the processing of the first eight FReDNet stations that provide sufficient long time series. The baselines between adjacent FReDNet stations are in the range of 20 to 60 km. ZOUF is part of the European Reference Frame Network (EUREF). One station (UDIN) has been uninstalled in 2006.

Each site is made of a GB-1000 Topcon receiver, an Ashtech Choke ring antenna, an affordable electric power supply system (solar panels, wind generators and backup battery), an effective data link system based on the modern Spread Spectrum technology (Figure 4 and 5). This configuration can yield GPS+GLONASS (GNSS) data both in real-time (for real time kinematic services) and for post processing purposes (RINEX files). GNSS data collected from each site is hourly and/or daily received and stored in the FReDNet database located in the main venue at CRS in Udine. All the GPS data coming from the remote sites are checked (through the standard quality check program QC from UNAVCO) and then published at the public ftp site ftp://www.crs.inogs.it/pub/gps/. Beside the standard ftp service, there is a new method to access the GPS data. It has been developed to help people that sometimes deal with GPS data like engineers, land-surveyors, architects. This kind of users usually do not want to spend a lot of time to pack an uncommon set of data, and usually they do not need the standard set of hourly or daily data which are available on the ftp site (most of the time used by scientists and researchers). A web interface lead the user to build up its own package in a very simple way. The new feature allows you to create a set of data that cover a fraction of an hour or more than an hour, with the preferred

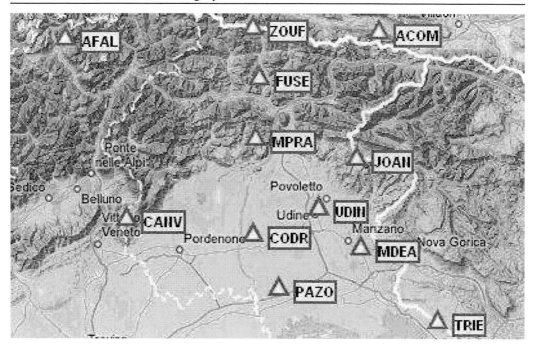

Figure 3. FReDNet GPS sites (red triangles).

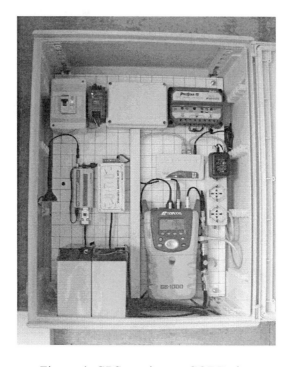

Figure 4. GPS receiver at CODR site.

Figure 5. GPS antenna used at the CODR site.

sample rate (1s, 2s, 4s, 4s, 5s,) and from one or more sites (see Figure 6).

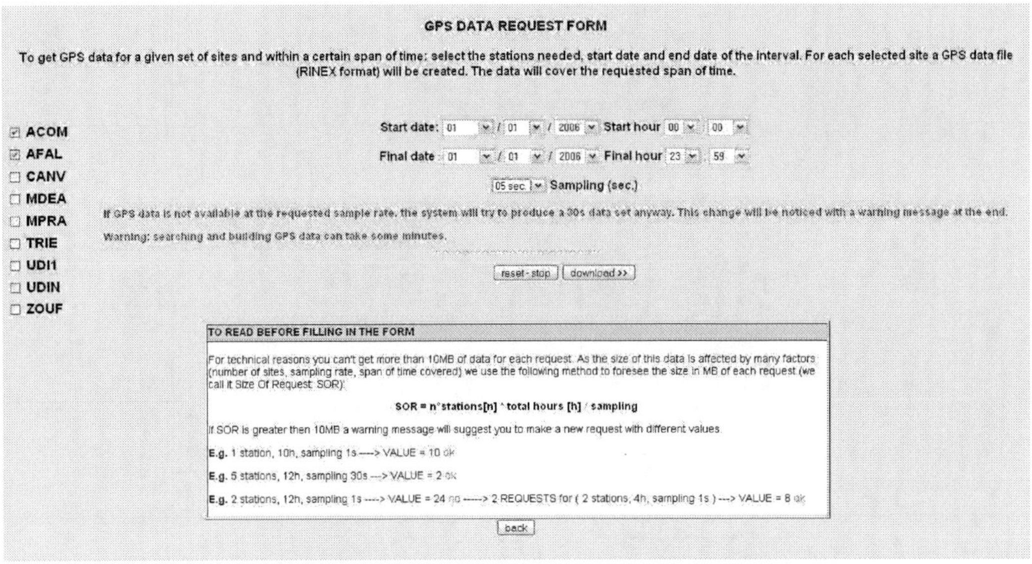

Figure 6. New request form for accessing the GPS database of FReDNet.

Since autumn 2006 CRS is involved in a project that regards real-time positioning based on the GPS technology. Based on a GNSS (GPS+GLONASS) infrastructure of permanent sites, CRS is going to enable a real-time positioning service with the reliable precision and accuracy needed for both cartographic and cadastral applications. A further target is the evaluation of the regional tectonic mechanisms and the monitoring of the active tectonics in some particular areas of the region to better assess seismic hazards. For that CRS has set up a local GPS campaign in the Tolmezzo area in 2006 (Figure 7) to provide a detailed picture of the local deformation between MPRA and ZOUF, an area of supposed seismic

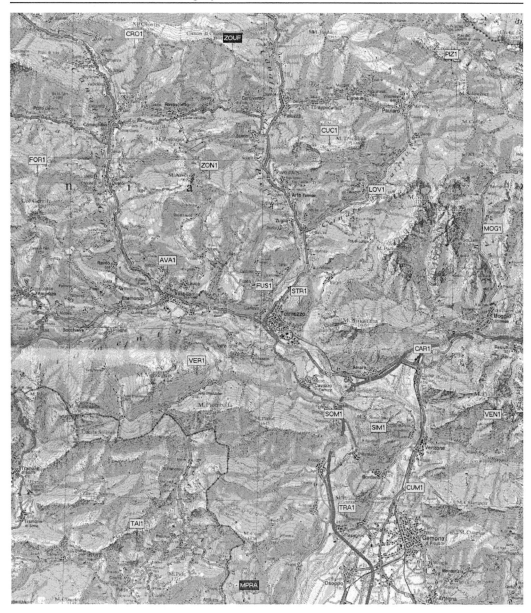

Figure 7. Site map of the Tolmezzo campaign. Yellow tags are the GPS campaign sites, blue tags are the GPS permanent sites of FReDNet.

gaps as well as location of the severe 1976 seismic sequence. The GPS campaign involved to establish a local benchmark network (Figure 8) and to survey the benchmarks with the suitable facilities (Figure 9). Currently (September 2008), the second campaign is taking place in the Tolmezzo area.

Figure 8. Benchmark installing.

Figure 9. Tripod and GPS facilities during a benchmark survey of the Tolmezzo campaign.

2.2. Data Acquisition

To investigate the deformation, we processed continuous GPS data acquired from June 2002 to November 2006. All GPS observations and ancillary data used in this analysis are public domain, collected from global and European continuous GPS networks (i.e. EUREF, IGS). For each day of the analysis interval, measurement files in compressed Receiver Independent Exchange (RINEX) format were downloaded for all the sites included in the analysis. Ancillary data required to analyze the GPS measurements, including broadcast navigation files and very precise orbits, were retrieved from the SOPAC archive (Scripps Orbit and Permanent Array Center; http://sopac.ucsd.edu).

2.3. Analysis Method

We processed the GPS observations using the *GAMIT/GLOBK* GPS analysis package developed at the Massachusetts Institute of Technology (MIT) and Scripps (http://web.

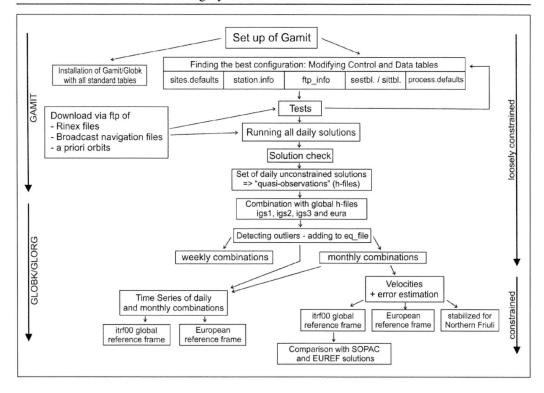

Figure 10. Overview of the continuous GPS processing.

mit.edu).

The *GAMIT* modules estimate the three-dimensional relative position of the GPS ground stations and satellite orbits. The primary output of *GAMIT* is a loosely-constrained solution file (*h-file*) of estimates and an associated covariance matrix of station positions and (optionally) orbital and Earth-rotation parameters ('quasi-observations'). In this solution, neither the coordinates of the tracking sites nor the GPS satellites orbits are tightly constraint, i.e. are not fixed in a well-determined reference frame, but baseline lengths are determined very precisely in the loosely-constrained solutions and the entire GPS network and GPS constellation can be rotated and translated as a rigid body [Larson et al. 1997].

The 'quasi-observations' are input to *GLOBK* (or other similar programs, e.g. QOCA) for combination with global *h-files* to estimate station positions and velocities, as well as orbital and Earth-rotation parameters.

Finally, the loosely-constrained solutions were transformed into a consistent reference frame using *GLORG* to derive deformation velocities. The choice of the reference frame, defining the origin, scale and orientation of the geodetic coordinates [Larson et al. 1997], depends on the needs of the study.

Figure 10 gives an overview of our GPS processing scheme.

2.3.1. Setting up *GAMIT*

The first step in analyzing GPS data is organizing the data in such a way that it can be handled efficiently and without errors by the processing programs. We made decisions about how many days to analyze, what stations should be included, and what are the best survey-specific parameters during this process.

The analysis software *GAMIT* is composed of distinct modules (for detailed information about the modules see the GAMIT/GLOBK manual). For processing large numbers of days, significant effort can be saved using the automatic script *sh_gamit* that runs the modules one after the other. The major preparation effort required is setting up the control and data files (or tables). These files have to be modified according to the requirements of the particular study area, while all standard control and data files need to be linked or copied into the experiment directory. Most tables are included in the *GAMIT* software package, e.g. 'Standard' tables to provide lunar/solar ephemeredes, the Earth's rotation, geodetic datum, and spacecraft and instrumentation information.

Control and Data Files

a priori **coordinates and velocities:** There are two files of *a priori* coordinates and velocities used by the *GAMIT* modules. The file ending in *.apr* (set as aprf in *process.defaults*) contains the Cartesian coordinates (position and velocity) of stations that are left unchanged throughout the processing. The *L-file* (*lfile.*) is updated by *sh_gamit* at the end of the daily processing if adjustments exceed a specified value (here: 0.3 m).

sites.defaults: This file controls the use of the stations in the processing. It specifies which local and IGS [International GPS Service; Beutler et al. 1994] stations are to be used and how station log-data are to be handled. The number of stations is determined by dimensions set during the *GAMIT/GLOBK* compilation, and can be tailored to fit the requirements and capabilities of the analyst's computational environment.

In our study, we chose the following permanent GPS sites for processing:

- 8 FReDNet stations;

- 26 stations around the Adriatic;

- 8 stations as reference for stable Eurasia.

station.info: The site-specific receiver and antenna information is recorded in *station.info*. The file includes a complete history of antenna and receiver hardware and firmware changes, and antenna heights for all sites. Accuracy of the database contents is crucial to the analysis, since errors or omissions in site histories, e.g. incorrect antenna height, will lead to incorrect modeling of phase observables and to offsets in the coordinate time series. We created the file using the most updated file from SOPAC, so the file is current for all IGS stations in the analysis. For the stations BOLG, COMO, FATA, GSR1, MSEL and ZOUF *station.info* was modified using the information on the log-files.

sestbl. and *sittbl.*: The *sestbl.* and *sittbl.* files are the two control files for the analysis. In these two files several parameters can be modified with the goal to optimize the configuration for data processing. We performed several numerical experiments to determine an 'optimal' configuration for the processing of the FReDNet GPS data. These experiments were based on combinations of the *sestbl.* and *sittbl.* files between the standard *sestbl.* and *sittbl.* tables from *GAMIT 10.21*, and the ones available on SOPAC (*sestbl.global*, *sestbl.regional* and *sittbl.refined*).

The following values were compared for the stations ACOM, MPRA, TRIE and ZOUF in the print output of the *Q-file* to determine the optimal solution:

- Postfit nrms, i.e. the square root of χ^2 per degree of freedom (at least < 0.25)

- Formal error

- Fract (at least < 10)

- Baseline uncertainties (at least $< 10\,\mathrm{mm}$)

Considering all the four criteria above, the best solution resulted from the combination of *sittbl.refined* both with *sestbl.* (here: Configuration 1) (Standard of *GAMIT release 10.21*) or *sestbl.global* (here: Configuration 2).

Configuration 1 results in smaller formal errors and baseline uncertainties. Furthermore, the 'Postfit nrms' is lower than the one of Configuration 2. In contrast Configuration 2 has lower 'Fract' values, while Configuration 2 often exceeds the acceptable value of 10. To benefit from the advantages of both configurations we created *sestbl.FREDNET* by a combination of *sestbl.global* and *sestbl.* (Standard *GAMIT* file).

In the paragraphs below, we provide a description of the most important parameters of *sestbl.FREDNET*:

Choice of Observable: *LC_AUTCLN* or *LC_HELP*?   *LC_AUTCLN* as Choice of Observable in *sestbl.* is a new feature of *release 10.2*. The *LC_AUTCLN* observable is the ionosphere-free linear combination (LC). The regular *SOLVE* solution uses the assignment of ambiguity parameters (site/sat combinations) and the resolution of wide-lane ambiguities made by *AUTCLN* using the pseudo-range. With modern receivers (e.g. as the ones of FReDNet), which are able to detect the carrier phase of the P-code for both the L1 and L2 ranges, it is recommended to use the new scheme *LC_AUTCLN* because it is more efficient at fixing bias and resolving phase ambiguities [R.W. King – MIT, and E. Serpelloni – INGV, personal communication 2006]. If *LC_AUTCLN* is used, the *LC_AUTCLN* command for *AUTCLN* wide-lane resolution must be present in the *autcln.cmd* file.

For a successful application, the algorithm *LC_AUTCLN* requires precise and unbiased pseudo-ranges. It has been recognized for several years that there exists a pseudo-range bias between the C/A and P1 for each satellite, the so-called 'Differential Code Bias' (DCB). Since April 2000, the Center for Orbit Determination in Europe (CODE; http://www.aiub.unibe.ch) provides DCB values for all active satellites on a monthly basis using the global IGS network. These corrections are applied automatically in *GAMIT 10.2* using the table *dcb.dat*. For data prior to the availability of satellite DCB's, the choice of *LC_HELP* as observable gives a more conservative approach in the use of ionospheric constraints and is recommended to be used [R.W. King, personal communication 2006]. With modern data, R.W. King [personal communication 2006] strongly suggests the use of *LC_AUTCLN* as observable, unless an inversion error is encountered in *SOLVE*.

As FReDNet was installed in 2002, we decided to process the data with *LC_AUTCLN*. Inversion errors occurred only for 20 of 1600 days processed.

Zenith delay For networks with an extension greater than a few kilometers, the zenith delay parameters should be estimated for each station (Zenith Delay Estimation = YES). For FReDNet, we set an interval of two hours for the calculation of a new value (Interval Zen = 2). The zenith delay *a priori* constraint was set to 0.5 m. To improve the definition of horizontal atmospheric gradients and to reduce their effect on the final position estimates, the elevation angle cutoff was set to 10°. The mapping function of Niell [1996], which describes how the average atmospheric paths delay varies as a function of elevation angle, was applied in the *sittbl*.

Antenna model The antenna (phase center) model was set to ELEV, to use an elevation-dependent model for the effective phase center of the receiver antenna. The elevation-dependent models are currently developed by NOAA (http://www.ngs.noaa.gov/) for the most commonly-used antennas and are included in the *GAMIT* files *antmod.dat* and *rcvant.dat*.

Tide model All sites are corrected for ocean tidal loading using the IERS03 ocean tide model. By setting 'Tides applied' = 31, the diurnal earth, semi-diurnal Earth, pole and ocean tides are applied. The fifth bit removes the mean for the pole tide.

Atmospheric loading The fact that atmospheric pressure loading (ATML) causes deformation of the Earth's surface has long been acknowledged and detected in GPS time series [Tregoning and van Dam 2005, and references therein]. The application of atmospheric loading improves of about 20 % of the scattering (wrms) of the vertical component of the solution. The improvement of the horizontal component is minimal. The atmospheric loading was not included in the processing of the FReDNet data (Apply Atm loading = N) because the ATML grid files are not part of the standard *GAMIT* tables and were not available after November 2005 [they are produced by the European Center for Geodynamics and Seismology; Tregoning and van Dam 2005]. The latest release of *GAMIT* (*10.3* – December 2006) fully supports this model capability, so we suggest evaluating the advantage of the application of ATML in the processing of the FReDNet data in future work.

2.3.2. Running *GAMIT*

About 1 600 days were successfully processed. The 'Postfit nrms' of the solution was unacceptably high (see below), or the process crashed and the problem could not be fixed, for only 29 days.

The most common FATAL message was an inversion error in *SOLVE*, due to problems the *LC_AUTCLN* option has with some data [R.W. King, personal communication 2006]. This software specific problem will probably be fixed in the future. We were able to complete the daily processing by changing the 'Choice of Observable' to *LC_HELP*, but we decided, for consistency, to exclude these solutions. Tests with a couple of other days showed that removing the station AQUI for specific runs led to a successful completion of the processing, even when using *LC_AUTCLN*. But an inspection of the observation file of AQUI did not give information about the error source.

2.3.3. Checking the Daily Solutions

We checked all the daily solutions to verify that they satisfied the following criteria:

- The 'Postfit nrms', i.e. the square root of χ^2 per degree of freedom, was smaller than 0.25. A value larger than 0.5 means that there are cycle slips that were not removed or associated with extra bias parameters.

- The 'Fract' value was less than 10 for all of the stations.

- The baseline uncertainty was less than 10 mm for most of the baselines

The 'Postfit nrms' of the solutions commonly ranges from 0.16 to 0.18. In ca. 10 solutions the 'Postfit nrms' of 0.25 was exceeded. These daily solutions were not used in *GLOBK* and *GLORG*.

2.3.4. Combination with Global *h-files*

For the next processing steps the cleaned set of unconstrained daily solutions was passed to *GLOBK*. *GLOBK* is a Kalman filter that allows combining solutions from the processing of primary data from space geodetic observations. Developed as an interface with *GAMIT*, it accepts the output data of *GAMIT*, the 'quasi-observations' (*h-files*). These primary solutions must be a robust input, because it is not possible to resolve phase ambiguities with *GLOBK*. Furthermore, *GLOBK* cannot correct deficiencies in the primary phase analysis due to missed cycle slips or 'bad' data. The basic algorithms and a description of Kalman filtering are summarized by Herring et al. [1990] and Dong et al. [1998]. Similar to *GAMIT*, *GLOBK* is an ensemble of programs (for details see GLOBK manual).

The combination of solutions from the regional analysis with the ones from a global analysis is a robust approach to aid orbit determination and to strengthen the reference frame [McClusky et al. 2000, Herring 2003a].

For each day, we combined the solutions of the regional analysis, consisting of about 42 stations distributed around the Adriatic and on stable Europe, with the quasi-observations of about 210 GPS stations from a global analysis of IGS data performed by SOPAC (http://garner.ucsd.edu). Ten stations are both included in our regional analysis as well as in the global one by SOPAC, and therefore provide a link between the two networks (i.e. ARTU, BOR1, GLSV, JOZE, LAMA, MEDI, POTS, TIXI, VENE, WTZR).

The combinations were performed by applying the *GLOBK* programs. We performed a rescaling of the *h-files* using the prefit χ^2 from an initial combination. Although the nrms scatter of the double-difference residuals from the *SOLVE* run is passed to the *h-file*, this information is not used in *GLOBK*. This means that the covariance matrix from *SOLVE* is not scaled automatically during the processing.

The main steps performed are:

132   M. Bechtold, D. Zuliani, P. Fabris et al.

- Determination of the prefit χ^2 for each network and month by performing monthly combinations for each network (igs, eura and the regional analysis). These values are used to scale the covariance of the *h-files* of each network in the daily combinations;

- Performance of two runs of *GLOBK* for daily combinations. In the second run the daily *h-files* are rescaled by the prefit χ^2 of the first run. The results have shown that one re-run is sufficient to obtain a prefit χ^2 that is close to 1;

- Archiving the solutions and cleaning the working directory.

The aim of the manual rescaling is to adjust the scaling on the *h-files* such that the prefit χ^2 is close to 1. The prefit χ^2 (goodness of fit) provides a measure of the misfit of data relative to their formal uncertainties. A prefit χ^2 that is close to 1 results in one-σ data uncertainties that are equal to the average scatter of the data residuals about the solution [McClusky et al. 2000]. The data scatter provides an unbiased estimate of the uncertainties if the error spectrum is white and the data are spatially- and temporally-sampled homogeneously.

As ambiguities are reliable resolved for both the regional analysis as well as the global *h-files*, we used for further processing the biases-fixed with loose constraints binary *h-files*, following the suggestions of Herring [2003a; pp. 7].

Preparing the daily and monthly command files: The daily and monthly command files control the *GLOBK* run and specify:

- the scratch files used by the program, and the files containing within *a priori* station positions, satellite ephemeris information and earthquake/rename commands;

- the output options, which can be used to tailor the output for a desired run;

- the uncertainties and the stochastic nature of the parameters in the solution.

In the following sections, we will present the most important command lines.

eq_file **Defining earthquakes and renaming stations**   The earthquake file command tells *GLOBK* how to handle earthquakes or other discontinuities ('jumps') in the stations' positions, so that they do not affect the long-term deformation rates. Several commands are available to define the characteristics of the major earthquakes and their effects on GPS stations [Herring 2003a]. In particular, the *rename* command allows to exclude 'bad' stations for specified time spans or to define recognized antenna offsets. We merged the standard *eq_file 'eq_file.global'* of *GLOBK* with an additional file *'scec_eq.v2.1'* (from MIT), which contains further *rename* commands for global stations detected by a processing of the MIT [R.W. King, personal communication 2006], and used this file as the basis for the processing.

apr_file **for station coordinates**   As in *GAMIT*, the *apr_file* in *GLOBK* provides *a priori* station coordinates and velocities. We used the standard file of ITRF positions and velocities *itrf00.apr* available in the *GAMIT/GLOBK* software. With *GLOBK 4.0* there is no need to add coordinates for new stations that are not included in the standard *apr_file*, since these are read from the binary *h-file*, written by *htgolb* from the *a priori* coordinates used by *SOLVE* and stored in the *GAMIT h-file*. Updated coordinates can later be extracted from the *GLOBK* or *GLORG* output files and used as a supplementary file. We used this procedure in the *GLOBK* runs for time series and velocities. In general, the reference frame of the *.apr* file is not important with respect to the later stabilization and reference frame definition in the *GLORG* run.

apr_file **for Earth-rotation parameters**   The command *in_pmu* allows the polar motion and UT1 (universal time version 1) series used in the analysis to be updated from the most up-to-date file, which is available on SOPAC and in the incremental updates of *GAMIT/GLOBK*. We applied *pmu.bull_b* of the incremental updates.

MAX_CHI (max χ^2 increment) (max prefit difference) (max rotation)   This option allows the automatic exclusion of a *h-file* from the solution, if the *h-file* contain corrupted data or a station, for which there are no reasonable *a priori* coordinates. It is possible to define tolerances on the:

- χ^2 increment, when a new *h-file* is combined in the solution;

- change of the *a priori* parameter values. If the prefit difference exceeds this limit, then the estimate for this parameter is not included, or in case of a station coordinate, all three coordinates are excluded. When the *a priori* information is well-known, this value can be set small;

- allowable pre-solution rotations of the prefit coordinates before they are compared with the current solution.

On the basis of several runs, we tightened the default values *'max_chi 100 10000 10000'* to *'max_chi 30 50 2000'*.

Selecting stations   For the daily combinations we did not exclude any stations.

Constraining parameters   With a variety of commands, it is possible to specify *a priori* σ for each parameter. The Kalman filter requires *a priori* constraints. If there is no constraint for a group of parameters, they will be ignored in the input binary *h-files*. As we combined the regional data of our analysis with global data, the best approach was to use the features of *GLORG* for the stabilization, i.e. the definition of the reference frame. With this approach, it was possible to allow the positions and velocities of GPS sites and satellites to be loose.

Another set of satellite parameters is for satellite antenna offsets. When the *a priori* values for satellite antenna offsets are not fixed, there is a 30 mm jump in height for five months at the end of 2005. At this period of time, SOPAC estimated the satellite antenna phase center position [T.A. Herring – MIT, personal communication 2006]. By freeing the parameters, there is a scale change that causes the jump in height. By fixing the antenna, offsets are constrained back to the standard values, in the case that they have been estimated.

As most of the information about polar motion and UT1 comes from the global GPS data themselves, the Earth orientation parameters (EOPs) are usually left relatively unconstrained, if combining with global *h-files* [Herring 2003a; pp. 40].

2.3.5. Outlier Detection

Although 'bad data' were excluded by previous constraints and solution checks (see above), an additional outlier detection, after having computed the daily computations, was necessary. For the outlier detection, we plotted the time series of all stations using *GLOBK* and *GLORG*. We used the following approach to detect the outliers:

Manual outlier detection: In a first manual inspection, we evaluated the time series for data with very large error bars. For eight days the error bars were very large (m-scaled) for all sites. For these days, the global *h-files* that are needed for a successful reference frame stabilization are partly missing. For these days, we removed the complete daily combinations. Furthermore, three stations have large error bars over a time span of more than one week or month. The stations are MDEA, ORID and ZOUF. For these time spans, we excluded the stations using the *rename* command in the *eq_file*.

Automatic outlier detection: After the manual outlier detection, we applied a script that performs an automatic outlier detection. The script performed the following steps:

1. Error bar outlier detection

    - For each station, the script calculated the median error bar of each component (N, E and U)
    - For all error bars that exceeded the median error bar by a defined factor (here: 2.5), the days were excluded for the corresponding stations.

2. Repeatability outlier detection

    - For each site the wrms of each component (N, E and U) was taken from the time series *VAL.** file.
    - If the offset with the tenth data point before and after a checked point (day) exceeded the wrms by a defined factor (here: 2.5), that day was excluded for the corresponding station.

2.3.6. Monthly Combinations

The combination of daily quasi-observations into weekly, monthly or annual averages is a method to account for white noise [McClusky et al. 2000, R.W. King and T.A. Herring, personal communication 2006]. R.W. King and T.A. Herring proposed that monthly combination provide sufficient observations for good statistics on the residuals about the linear trends. By scaling the daily combinations, such that the prefit χ^2 of each monthly combination is close to 1, this method essentially accounts for the white noise during velocity estimation (see section 'Determining the site velocities and estimating the error' for further reading on error estimation).

The monthly combinations were performed applying the script as carried out for the daily combinations. We included a rescaling of the daily combinations using the prefit χ^2 from an initial combination. We determined the prefit χ^2 for each day by performing a monthly combination. These values are used to scale the covariance of the daily combinations in a second *GLOBK* run for monthly combinations. The results have shown that one re-run is sufficient to obtain for each monthly combination a prefit χ^2 that is close to 1.

It should be noted that we appended the outliers detected by the outlier detection to the earthquake file used for the daily combinations, before we computed the monthly combinations. In this way, the monthly combinations are based on cleaned daily combinations.

2.3.7. Defining the Reference Frame

In *GLOBK* there are several possibilities for defining the reference frame used to estimate positions and velocities. The choice of the method mainly depends on whether the data set is small and only regional, or large, containing regional as well as global *h-files*. For analyses in which only regional data are included, the stability of the solution can be maintained by applying realistic *a priori* constraints, either on the orbits and Earth orientation parameters, or on the positions and velocities of a specific subset of stations.

The most robust approach is to base the reference frame stabilization on the combination with global *h-files* [Herring 2003a; pp. 37]. After all of the data were combined by *GLOBK* with loose constraints, *GLORG* allows the reference frame of the solution to be specified. The late imposing of constraints on the solutions is very useful, as different reference frames may be tested and realized very quickly. *GLORG* is the 'origin fixing' program for the data analysis. 'Origin fixing' means that translation, rotation and scale are estimated by a minimization of the deviations between horizontal positions and velocities given in the *a priori* station position file (**.apr*). The stations, for which the minimization is performed, are defined with the *stab_site* command in *GLORG*, or by a link to an external file. It should be noted that the *apr_file* used for *GLORG* does not need to be the one used for *GLOBK*. Focusing on our specific goals, we performed four different realizations of reference frames:

The ITRF00 global reference frame: The ITRF00 ('00' stands for the year 2000; sometimes also ITRF2000) is the most recent and most accurate realization of the International Earth Rotation Service (IERS) Terrestrial Reference System (ITRS) and has therefore the most reliable, precise and up-to-date datum definitions. It is a non-inertial reference system rigidly fixed to the Earth [Gatti and Stoppini 2000]. The origin of the XYZ Cartesian system is the Earth's center of mass. The Z-axis passes through the North pole. The X- and Y-axes lie on the equatorial plane. The international reference frame (ITRF00) is defined by a set of estimated coordinates of VLBI, SLR and GPS global stations, which form a 'fundamental polyhedron', and performs a three-dimensional network on the Earth's surface for determining the deformations of the crust. The reference epoch of the ITRF00 solution is 1997.0.

We performed the ITRF00 reference frame realization, because it is the most robust geodetic solution, and allows to compare our results with estimates by other analysis, e.g. SOPAC and EUREF.

The European reference frame of Altamimi et al. [2002]:   The international reference frame (ITRF) is not appropriate to study the Africa-Eurasia collision zone, including the movement of the Adriatic microplate. The resulting velocities of the ITRF are mainly characterized by the divergence/rifting of the mid-ocean ridge and features of the European deformation field are obscured. Therefore, there exist several European reference frames that are based on the approximation of a stable Eurasian crustal block to investigate the deformation around it.

GAMIT/GLOBK offers an *apr_file* that refers to an approximation of the stable Eurasian plate by Altamimi et al. [2002]. The movement of a continent, which is assumed to have a rigid-plate behavior, can be described as a rotation around a pole, the so-called Euler pole. Altamimi et al. [2002] used the ITRF00 velocity field to estimate the relative rotation poles for six major tectonic plates, e.g. stable Eurasia.

The European reference frame of McClusky et al. [2000]:   McClusky et al. [2000] determined an Eurasian reference frame by setting the *a priori* velocities of 14 European stations to zero and two Asian stations to a specific value. We followed this approach by creating an *apr_file* for the 14 European stations, taking the coordinates of the ITRF00 frame, and setting their velocities to zero. We did not take into account the two Asian stations, because the focus of our study is central Europe and not the eastern Mediterranean as in McClusky et al. [2000]. The advantage of this approach is its independence of specific files for *a priori* velocities of European GPS sites. The definition of the reference frame is very simple and non-ambiguous: The solution is stabilized for a set of 14 stations that are not considered to be moving. After the performance of the stabilization, the residual velocities of the 'stable' sites represent intra-plate deformation, e.g. the isostatic rebound of Scandinavia and Northern Germany.

The solution of Altamimi et al. [2002] is controversial as the stations used east of the Ural Mountains are all in, or close to, the central Asian deformation zone [Le Pichon and Kreemer 2004]. That is why we applied the approach by McClusky et al. [2000] for an alternative realization of an European reference frame.

The stabilization for northern Friuli: Because the northern GPS stations of FReDNet (AFAL, ACOM and ZOUF) are localized to the north of the main seismicity of the Friuli area, we decided to stabilize the solution for these stations to focus on the movement of the sites in the south. The result was used to estimate the angle of collision of the Adriatic plate, as well as the shortening rate within the Friuli region.

2.3.8. Plotting Time Series

The generation and evaluation of time series of station coordinates is an important tool to identify outliers and to study seasonal and semi-seasonal variations, as well as to receive an impression of the data quality. For the time series we ran *GLOBK* individually for each day including all stations. Using the daily *com* and *sol* files of the *GLOBK* run, we applied *GLORG* for the different realizations of the reference frames.

The *GLOBK cmd* file for time series (repeatabilities) corresponds to the one for the daily and monthly combinations. For the stabilization a *GLORG cmd* file is needed. In that file, the parameters for the stabilization are specified.

For the stabilization, i.e. the process of minimization of the deviations between horizontal positions and velocities given in the *a priori* station position file, *GLORG* is able to apply three operations: translation, rotation and scale. It is possible to restrict this process to only one or two operations. In general it is convenient to estimate only the translation, if the stabilization is performed for very few sites [Herring 2003a; pp. 43]. We followed this approach for the stabilization of northern Friuli. For the European and international reference frames we estimated both translation and rotation, as the *a priori* information is comprehensive enough. Because at the moment, it is not clear whether more accurate results are generated when scale changes are estimated [Herring 2003a; pp. 43], we did not include an estimation of the scale change.

With *GLORG* it is possible to define stations of the *stab_site* list, which define the reference frame, that are retained because of large σ values. We loosened the height tolerance to include all stations.

The coordinate system stabilization is controlled by three characteristics:

- The number of iterations

- The site-relative weight: A value of 0.0 signifies that all sites are weighted equally throughout all iterations. A value of 1.0 means that the site σ of the previous iteration are used to weight the sites in the subsequent iteration. A value between 0 and 1 allows only a partial weighting. Default is 0.5.

- The n-σ value: This value allows eliminating sites that are discordant with the *a priori* coordinates.

As the iterations run very fast, we increased the default value of two to four iterations. Concerning the value for the site relative weight, we applied two different approaches. For the European and international reference frames several sites are available for the stabilization process for each day. Therefore weighting the sites is convenient. For these reference frames, we set the site relative weight value to 0.8. For the stabilization for northern Friuli, we decided not to include weight, as only three stations are available. For n-σ, we adopted the default value of 4.

The output of *GLORG* are daily and monthly *org* files, which are used by *sh_plotcrd* to plot the time series.

2.3.9. Determining the Site Velocities and Estimating the Error

Instead of computing one *org* file for each day, week or month, it is also possible to run *GLOBK* and *GLORG* for the whole time span. This approach allows a realistic estimation of the site-specific errors of the velocities. Furthermore, one can include more restrictions concerning data used during the stabilization.

Bias in the velocity estimates: GPS coordinate time series have significant seasonally driven signals with an annually repeating component, which might bias velocity estimates [Blewitt and Lavallée 2002]. Generally, reduction of velocity uncertainty is associated with longer time series. The total time span of observations has an important influence on the uncertainty of site velocities from space geodetic data [Zhang et al. 1997, Mao et al. 1999].

A major component of annual signals is the physical monument motion [Blewitt and Lavallée 2002]. In most cases, the dominant reason is surface loading due to hydrology and atmospheric pressure [Blewitt and Lavallée 2002], but also anisotropic expansion of parts of the monument (e.g. the rock) due to temperature changes can have major effects at some sites. There is a controversial discussion concerning which time spans of data affect the velocity estimates less. Dixon and Mao [1997] suggested that the influence of annual errors on velocity estimates would be minimal for an integer number of years, but would have strong effects on velocity estimates for '2.5 year' time spans. Blewitt and Lavallée [2002] contradicted this statement and demonstrated that the bias drops rapidly towards integer-plus-half year data spans. As we deal with time spans that are all ≥ 2.5 years, according to Blewitt and Lavallée [2002] the largest velocity bias should be ca. 0.1 mm/yr for the '3.3 year' time span of station ACOM (the next time span closest to an integer year; see Table 1). As this discussion is controversial and the maximum bias of 0.1 mm/yr is rather small, we took full advantage of time series for velocity estimation and did not censor them to fulfill either of the two suggestions.

Table 1. Data time span of the eight FReDNet stations.

	Data time span in years
ACOM	3.3
AFAL	3.4
CANV	2.5
MDEA	3.8
MPRA	4.2
TRIE	3.7
UDIN	3.9
ZOUF	4.4

globk_vel.cmd: In the following, we present the *globk_vel.cmd* file. We only refer to features that are new or differ from the *cmd* files in terms of repeatability and combination.

The data can be processed separately in normal and reverse time order. In each case, the stations coordinates, Earth rotation parameters, and satellite orbit parameters refer to the first epoch. With this option, it is possible to check whether the forward and backward solutions are consistent.

It is possible to efficiently reduce the 'Forward χ^2 Increments' of the velocity solution by removing stations that are characterized by a large weighted rms (wrms). Using the output file *SUM.FREDNET.sort.wrms* of *sh_plotcrd*, we identified nine stations with large wrms. The limits that we defined to exclude three to four stations from each component were:

- for the N (north) component: wrms $> 6\,$mm; stations removed: KOKB, ZWEN, FAIR, YAKT;

- for the E (east) component: wrms > 9 mm; stations removed: KWJ1, MALD, KOKB;

- for the U (up) component: wrms $> 20\,$mm ; stations removed: ZIMM, QAQ1, DRAG.

It is possible to define an integer that specifies the minimum number of times a station must appear in the *h-files* to be included in the analysis. This option is useful as it is not convenient to stabilize for site coordinates and velocities that are only based on a few data points. We decided to use only sites that appear at least in 18 monthly combinations.

Estimating the error: It is crucial that the uncertainties in the velocity estimates are properly estimated. Formal errors underpredict the true scatter, or repeatability, of individual estimates. The error of velocities determined using GPS data has two main components [e.g. Miller et al. 2001]:

- non-correlated (also called white) noise

- temporally-correlated (random walk and flicker) noise.

Pure white noise models underestimate uncertainty [Johnson and Agnew 1995]. For a long time it was assumed that each length measurement is independent and that a normal distribution characterizes the measurement error. Langbein and Johnson [1997] disproved this and demonstrated the existence of temporally-correlated error in time series.

Time-correlated noise effects include potential monument motion unrelated to crustal deformation [modeled as a Brownian motion, Langbein and Johnson 1997], uncertainty in the satellite orbit parameters, and atmospheric and local environmental effects [Mao et al. 1999]. The time-correlated noise consists of two components: flicker and random walk noise. Both are special cases of power law noise. In power law noise, the power density function is related to the frequency, f, by:

$$P_0/f^n, \tag{1}$$

where P_0 is a constant and n is the spectral index.

Thus flicker noise is defined when $n = 1$ and random walk is defined when $n = 2$ [Langbein 2004].

It is an ongoing discussion whether flicker or random walk noise is more consistent with temporal correlations in a GPS time series [Langbein 2004, and references therein]. At the moment it is only possible to implement a random walk noise in *GLOBK*. To account for flicker noise it is necessary to perform a separate spectral analysis outside *GAMIT/GLOBK* [see e.g. Nikolaidis 2002]. For our error estimation we concentrated on a combination of white and random walk noise, an approach applied in several publications [e.g., McClusky et al. 2000, R.W. King, personal communication 2006].

Langbein and Johnson [1997] proposed that random walk noise typically ranges from 0.5 to 2.0 mm/$\sqrt{yr}$. In general, random walk noise estimates largely rely on *a priori* assumptions about monument behavior that may not be well determined for a wide variety of geodetic monuments in realistic settings [Miller et al. 2001]. FReDNet stations are set up on different rocks (limestone, porphyrite and loose quaternary sediments); in addition, different atmospheric effects are likely across the network that spans heights from ca. 100 m on the Venetian plain to ca. 2000 m in the Southern Alps. The variety of settings probably causes random walk effects that differ largely from site to site.

When continuous data are available, it is possible to estimate a random walk component separately for each station using the 'realistic sigma' algorithm of Herring [2003b; e.g., Reilinger et al. 2006]. It can be applied by the command *sh_gen_stats*, an additional shell-script within the *GAMIT/GLOBK* software. It uses the *VAL* file of the time series and provides as output a file *rw*.rw* of *mar_neu* commands, which define the random walk component of each site and can be put into *GLOBK*. Table 2 presents the random walk components calculated for the FReDNet stations.

Table 2. Site-specific estimated random walk for each component (N, E and U) of the FReDNet stations.

	random walk components* (in mm/$\sqrt{yr}$)		
	N	E	U
ACOM	0.11	0.11	3.93
AFAL	0.08	4.18	4.74
CANV	1.05	0.33	5.54
MDEA	0.17	0.10	4.22
MPRA	0.17	0.20	4.78
TRIE	0.16	0.36	3.84
UDIN	0.83	0.87	7.89
ZOUF	0.12	0.28	4.63
Average	0.34	0.80	4.95

* estimated by the 'realistic sigma' algorithm

It should be noted that the random walk is time dependent. Random walk has a smaller influence at the beginning of the time series, causing the estimated velocities to differ from the slopes in the time series. The estimated velocities with random walk are supposed to provide more realistic values than those inferred from the slopes of the time series [R.W. King, personal communication 2006]. As the random walk is time dependent, forward and backward solutions differ. In this analysis the differences were in the range of ca. 0.1 mm for the ITRF00 and the Altamimi et al. [2002] reference frame, small values for which the forward solution should be accepted as final solution [R.W. King, personal communication 2006]. In the McClusky et al. [2000] reference frame and the stabilization for northern Friuli the forward and backward solutions differed significantly (0.5 to 2 mm/yr). The large differences are caused by the implementation of a random walk effect on stations whose *a priori* velocity is zero. Furthermore, the *a priori* information is less comprehensive (less stations), which results in a reference frame that is weaker and more vulnerable for the application of a random walk effect. To fix the forward and backward solution, we set the random walk components to zero for all sites that are used in the McClusky et al. [2000] reference frame and the stabilization for northern Friuli. This approach was adequate, as we were not interested in weakening our reference frame with the implementation of a random walk effect, but to estimate the random walk included velocities and errors for all the other sites.

Rescaling the solution: We used the average of the 'Forward χ^2' and the 'Backward χ^2' for each month of the velocity solution to scale the position variances of each monthly combination, so that the reduced χ^2 statistic of the velocity solution was approximately 1 for the next run. The final rescaled solution properly accounts for the white noise [McClusky et al. 2000].

Evaluating the stabilization: The rms of the velocity residuals of the 67 IGS core stations used for the ITRF00 reference frame was 1.12 mm/yr. Values ranging from 1 to 2 mm/yr signify successful stabilization [Herring 2003a; pp. 58]. For the Altamimi et al. [2002] reference frame realization, it resulted in 1.3 mm/yr, a similar value (66 IGS core stations used in stabilization). For the stabilization after McClusky et al. [2000] and for the stabilization for northern Friuli the values were 0.6 mm/yr and 0.14 mm/yr, respectively. The values are very low, indicating an even better stabilization, affected by using only 10 and 3 stations for stabilization, respectively.

2.3.10. Summary of GPS Processing

We consistently analyzed the whole data set using a three-step approach described by Feigl et al. [1993], Oral [1994] and Dong et al. [1998]. This approach is widely applied in GPS geodesy [e.g. McClusky et al. 2000, Serpelloni et al. 2007]:

- In the first step, we used daily GPS phase observations to estimate station coordinates, atmospheric zenith delay, orbital and Earth orientation parameters by means of the *GAMIT* software, while applying loose constraints to geodetic parameters.

- In the second step, we used *GLOBK* to combine, on a daily basis, the loosely-constrained solutions with the global and European loosely-constrained solutions

Table 3. Amplitude (in mm) of annual variations in the time series of the FReDNet stations.

	N	E	U
ACOM	1	1	8
AFAL	1	6	10
CANV	3	2	9
MDEA	1	1	10
MPRA	3	3	10
TRIE	2	2	10
UDIN	5	2	12
ZOUF	2	2	10
Average	2	2	10

(igs1, igs2, igs3 and eura) computed by SOPAC. In this step, both orbital and common station parameters were estimated for forward analysis with loose constraints. To account for white noise we combined the daily combinations into monthly averages, rescaling the daily *h-files* such that the final χ^2 is approximately 1.

- In the third step, we defined four different reference frames by minimizing the deviations between horizontal positions and velocities given in an *a priori* station position file (**.apr*) for a specific set of GPS sites using *GLORG*. All eight stations of the Friuli Regional Deformation Network (FReDNet), which form the focus of this chapter, present at least 2.5 years of data, the minimum time span for a reliable determination of the deformation velocity [Blewitt and Lavallée 2002].

We used 93 % of the initial data set in the final solutions. Of the FReDNet stations, only 3 % of the initial data were excluded.

3. Results of GPS Processing

3.1. Time Series

In GPS time series, annual variations typically have amplitudes of 2 mm in the horizontal and 4 mm in the vertical components, with some sites with twice these amplitudes [Blewitt and Lavallée 2002]. In the time series of FReDNet (Figure 11 and 12), the average value of the horizontal components is 2 mm (Table 3). The strongest annual variations in the horizontal components can be detected in the E component of AFAL and in the N component of UDIN. Possible explanations for the anomalous strong variations could be an irregular satellite coverage due to a nearby mountain (AFAL) and an asymmetric (thermal) expansion of the monument (AFAL, UDIN).

Because since 2002 no major earthquake has occurred next to the GPS sites of FReD-Net, there are no coseismic earthquake movements in the time series. But there was one significant negative offset (ca. 2 cm) in the U component of UDIN in 2004. Since there

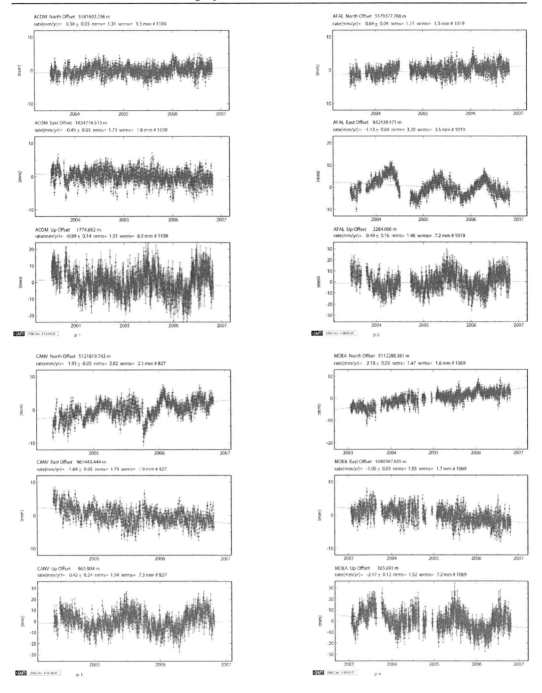

Figure 11. Daily time series in the Altamimi et al. [2002] European reference frame of four FReDNet stations (ACOM, AFAL, CANV and MDEA). Note that the scale of the X- and Y-axes changes.

were no changes in the instrumentation, this offset could signify subsidence of the building that is founded on Quaternary alluvial gravel.

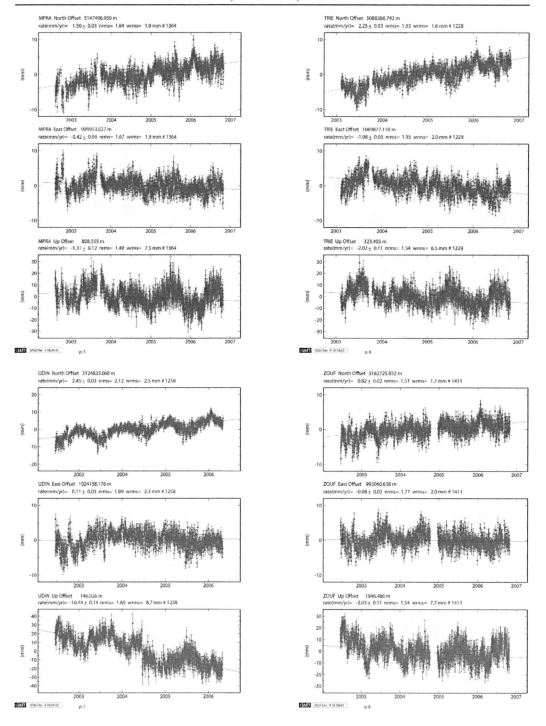

Figure 12. Daily time series in the Altamimi et al. [2002] European reference frame of four FReDNet stations (MPRA, TRIE, UDIN and ZOUF). Note that the scale of the X- and Y-axes changes.

Table 4. Velocities (in mm/yr) and their 1 standard deviation (66 % confidence level) of the N, E and U components of the FReDNet stations in the ITRF00 and European [Altamimi et al. 2002 and McClusky et al. 2000] reference frames as well as for the stabilization for northern Friuli.

	ITRF00			Stable Europe [Altamimi et al. 2002]			Stable Europe [McClusky et al. 2000]			Northern Friuli		
	N	E	U	N	E	U	N	E	U	N	E	U
ACOM	14.74±0.22	20.83±0.22	-0.58±1.23	0.66±0.22	-0.37±0.22	-0.48±1.23	0.67±0.21	0.14±0.21	-1.13±1.23	-0.15±0.39	0.34±0.05	1.69±0.34
AFAL	15.09±0.21	20.52±1.15	0.45±1.36	0.87±0.21	-0.44±1.15	0.55±1.36	0.92±0.20	0.05±1.14	0.10±1.36	-0.03±0.44	-0.57±0.06	2.48±0.37
CANV	17.01±0.69	19.10±0.41	-0.51±1.73	2.83±0.69	-2.04±0.41	-0.41±1.73	2.86±0.69	-1.56±0.41	-1.17±1.73	1.95±0.74	-1.19±0.63	2.23±1.73
MDEA	16.33±0.25	19.96±0.21	-2.36±1.18	2.25±0.25	-1.37±0.21	-2.26±1.18	2.25±0.24	-0.87±0.19	-2.97±1.18	1.47±0.41	-0.64±0.59	0.61±1.19
MPRA	15.89±0.24	20.44±0.25	-1.14±1.18	1.76±0.24	-0.74±0.25	-1.04±1.18	1.79±0.23	-0.26±0.24	-1.71±1.19	0.91±0.24	-0.04±0.36	1.95±1.19
UDIN	17.32±0.49	21.03±0.50	-8.05±1.55	3.23±0.49	-0.24±0.50	-7.95±1.55	3.25±0.49	0.23±0.49	-8.56±1.56	2.41±0.54	0.42±0.67	-4.89±1.56
TRIE	16.34±0.24	20.60±0.34	-2.27±1.12	2.30±0.24	-0.84±0.34	2.17±1.12	2.29±0.23	-0.35±0.33	-2.29±1.13	1.56±0.59	-0.08±0.82	0.67±1.13
ZOUF	15.18±0.20	20.71±0.28	-2.59±1.13	1.05±0.20	-0.39±0.28	-2.49±1.13	1.07±0.19	0.09±0.27	-3.13±1.14	0.17±0.08	0.23±0.05	0.98±0.31

3.2. Site Velocities and Error

In Figures 13 and 14 the velocities of the Adriatic and the eight FReDNet sites are represented by vectors. The error ellipses indicate the 95% confidence level (2 standard deviation). In Figure 14, the map additionally shows the structural map by Galadini et al. [2005] and the DEM of the Friuli area (Data source: http://www2.jpl.nasa.gov/srtm/).

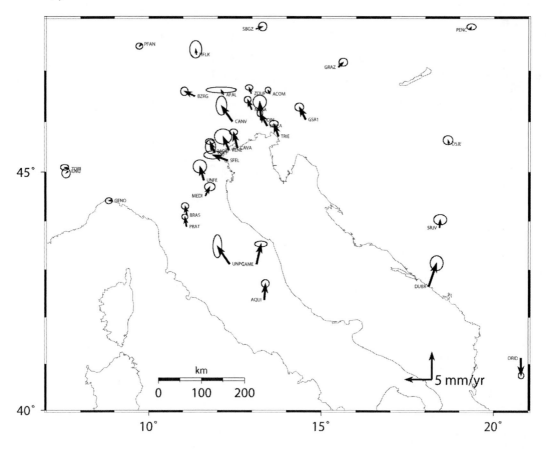

Figure 13. Velocities of the Adriatic sites in the European reference frame of Altamimi et al. [2002].

The velocity solutions for the two European reference frames differ slightly (see Table 4). The absolute velocities are very similar, whereas the movement direction with respect to stable Europe is different. In the reference frame of Altamimi et al. [2002] all station velocities show a westward swift of about 10 to 20° with respect to the reference frame of McClusky et al. [2000].

The difference between the two European reference frame realizations is caused by two factors:

- The reference frame of Altamimi et al. [2002] is partially based on stations located east of the Ural Mountains; stations that are probably affected by the Asian deformation zone. In the McClusky et al. [2000] European reference frame realization of the

Global Positioning System Constraints on Plate Kinematics... 147

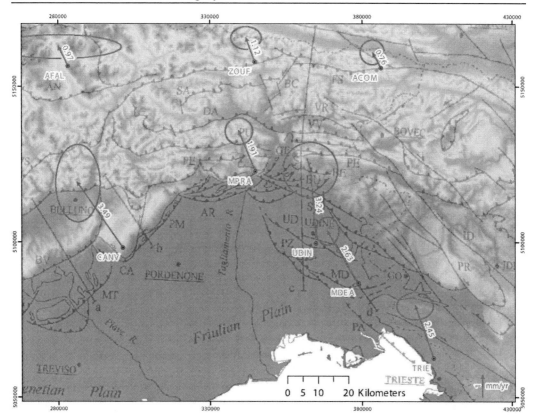

Figure 14. Velocities in the European reference frame of Altamimi et al. [2002] [Bechtold et al. submitted].

present chapter we did not include Asian sites. Consequently the resultant velocity field is not influenced by internal deformation in Asia.

- Altamimi et al. [2002] and McClusky et al. [2000] did not choose exactly the same sites as approximation of stable Europe. Thus, part of the difference between the two reference frames could be also based on the different selection of the stable European sites.

The difference between the two European reference frame realizations demonstrate the necessity of a development of a standard European reference frame for GPS geodesy, similar to the ITRF00 for global studies. Currently, published European velocity fields are hard to compare due to different reference frame realizations.

The velocity field of the Adriatic region (Figure 13) shows central Adriatic stations (CAME, DUBR) moving N to NNE and northern Adriatic stations moving NNW with respect to stable Europe after Altamimi et al. [2002]. This observation is consistent with the anti-clockwise rotation of the Adriatic microplate, proposed by Anderson and Jackson [1987]. Furthermore, Figure 13 demonstrates the particular tectonic position of the Friuli area situated between slightly- to not-moving sites of the Alps (e.g. PFAN, HFLK, SBGZ) and NNW-moving sites of Northern Adria. A detailed view of the Friuli velocity field

(Figure 14) shows that the FReDNet GPS sites can be divided into three groups:

- AFAL, ACOM and ZOUF (northern Friuli): With respect to stable Europe, the velocities range from 0.7 to 1.1 mm/yr. The movement direction is N to NNW.

- CANV, MDEA, TRIE and UDIN (southern Friuli): With respect to stable Europe, the velocities range from 2.3 to 3.5 mm/yr. The movement direction is N to NNW.

- MPRA: With ca. 1.9 mm/yr, the horizontal velocity of MPRA lies between those of northern and southern Friuli. MPRA probably represents an intermediate position.

The angle of collision of southern Friuli with respect to northern Friuli can be inferred from the solution stabilized for northern Friuli. Less weight should be given to the movement direction of UDIN and TRIE, as both are on monuments on top of buildings; it cannot be assured that their movement is completely tectonic. The movement direction of CANV and MDEA suggests a NNW movement at 328 to 335°. The velocity field indicates shortening within the Friuli area of about 1.6 to 2.2 mm/yr. The uncertainties in the U component are too large for reliable conclusions.

3.3. Comparison of the Results with SOPAC and EUREF Solutions

The GPS velocities published by different institutions always present slight differences. A 'correct solution' does not exist. To evaluate whether the GPS velocities resulting from our study are reasonable, we compared the results with those of SOPAC and EUREF (Table 5). SOPAC and EUREF, which both are major participants in the International GPS Service (IGS), provide weekly-updated velocity solutions based on the model of Nikolaidis [2002] and the work of the EUREF Permanent Network (EPN), respectively. The difference between the SOPAC and EUREF solutions is an indicator for the variations of official velocity solutions. The average difference between the horizontal velocities is ca. 0.5 mm/yr, and the one between the vertical velocities is ca. 1 mm/yr. The average difference between the station coordinates ranges from 0.004 to 0.007 m. The average differences between the velocities of our study and the ones of SOPAC and EUREF range from 0.27 to 0.84 for the horizontal and from 1.33 to 1.56 mm/yr for the vertical component. The average differences for the station coordinates range from 0.004 to 0.012 m. The differences between the SOPAC and EUREF solutions and those of our study are of the same order. The comparison supports the velocity solutions of our study, which therefore can be considered as reliable as those published by these institutions.

4. Discussion of Velocity Field

Since the FReDNet stations have limited spatial distribution and the tectonics of the Friuli area are so complex, it is not possible to directly determine active structures from the GPS velocity field. Due to the long distances between two stations, differential movements could always be related to more than one fault. But the resultant GPS velocity field can at least

Table 5. Comparison between the solutions of this analysis (FREDNET) and the ones published by SOPAC and EUREF, in the ITRF00 reference frame.

Velocities	Difference in mm/yr		
	N	E	U
SOPAC – EUREF	0.61	0.33	1.14
FREDNET – SOPAC	0.27	0.66	1.33
FREDNET – EUREF	0.43	0.84	1.56
Coordinates	Difference in m		
	X	Y	Z
SOPAC – EUREF	0.004	0.005	0.007
FREDNET – SOPAC	0.009	0.004	0.012
FREDNET – EUREF	0.012	0.005	0.012

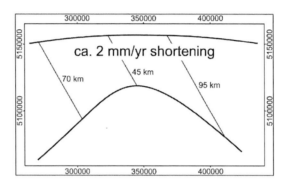

Figure 15. The GPS velocity field constrains the area, in which the shortening of ca. 2 mm/yr takes place. In the central sector the length is about half of the distance in the west and east.

constrain the area in which the shortening of ca. 2 mm/yr takes place (Figure 15). The southern border is defined by the stations CANV, UDIN, MDEA and TRIE, presenting deformation velocities that are in the range of those supposed for the Adria. The stations ACOM, ZOUF and AFAL define the northern border of the shortening sector. Those stations are north of the active seismicity area and demonstrate only a residual northward movement of about 1 mm/yr (with respect to stable Europe). The intermediate station MPRA indicates that along two or three thrust faults, 5 to 10 km south of the station, shortening of ca. 1 mm/yr occurred during the last four years (see e.g. Figure 14). The area of shortening (Figure 15) defined by the GPS velocities indicates that shortening of ca. 2 mm/yr takes place over a length of ca. 45 km in the central sector, whereas the same amount of shortening occurs over up to double the length in the west and east of it. This fact alone could contribute to the higher density of seismic events in the central sector during the last 30 years. More GPS stations are needed however within the area of shortening to better define its borders.

4.1. Influence on Velocities due to Strain Accumulation along Faults

Displacements on faults accommodate a substantial fraction of the strain occurring in the upper crust. Displacement on faults occurs continuously (fault creep), or abruptly during earthquakes [Turcotte and Schubert 2002]. Between earthquakes the fault remains locked. This is known as 'stick-slip behavior' [Turcotte and Schubert 2002]. An earthquake occurs when the stress across the fault builds up to a sufficient level to cause a rupture of the fault. During the elastic rebound, the elastic energy stored in the adjacent rock is partially dissipated as heat by friction on the fault and is partially radiated away as seismic energy [Turcotte and Schubert 2002].

A simple model of the earthquake cycle of Savage and Burford [1973] was the basis for many interpretations of geodetically-determined velocities and strain rates, along strike-slip faults, subduction zones, as well as intracontinental thrusts [e.g. Bennett et al. 1997, Gahalaut and Chander 1997, Le Pichon et al. 1998; Meade and Hager 2005]. It describes the surface deformation velocities with increasing distance to a strike-slip fault for the interseismic and coseismic phase, and the long-term geologic behavior. The interseismic velocity profile is characterized by a smooth transition across the fault, in contrast to the coseismic and long-term geologic profiles, both of which are discontinuous. The block velocity is equal to the sum of the coseismic (slip deficit) and interseismic velocities.

The geodetically-determined velocities close to faults are thus composed of both an interseismic and coseismic component. For South California, an area characterized by high relative plate motion (ca. 50 mm/yr), Meade and Hager [2005] compared the moment release due to small events with the moment accumulation expected for larger events and noted that the displacements due to small earthquakes (Mw, moment magnitude < 5.0) are significantly smaller than the displacements associated with interseismic strain accumulation. Thus, except for the effects of a few large earthquakes, most surface deformation is associated with interseismic strain accumulation, not coseismic strain release. In the Friuli area, where the highest earthquake magnitudes are in the range of M = 6.5, and relative plate motions are only in the range of one or two mm/yr, coseismic strain releases of earthquakes with M < 5, which are hard to detect in working time series, should have a stronger influence on the observed GPS deformation velocities than in South California.

As already mentioned the GPS stations in the Friuli area are too scarce to define active structures; but under the assumption of certain active structures (e.g. as proposed in literature), GPS velocities can be evaluated if they fit the assumption. Differences between measured and supposed velocities may indicate strain accumulation along these faults.

At one FReDNet station (i.e. CANV - Caneva), the measured velocity indicates strain accumulation. The station CANV is situated ca. 4 km northwest of the Polcenigo-Maniago Thrust, which is considered as recently-active, e.g. responsible for the 1873 (M > 6) earthquake [Galadini et al. 2005]. The long-term geologic (block) velocity near CANV should therefore be significantly lower than the one of the Adria, similar to the intermediate velocity of MPRA. The geodetically-determined deformation velocity of CANV is higher than the supposed block velocity, and in the range of the ones of UDIN, MDEA and TRIE. Figure 16 illustrates that the currently-observed velocity is mainly characterized by the interseismic component and probably related to strain accumulation along the Polcenigo-Maniago Thrust. A differential horizontal movement of 1 mm/yr and a locking period of

100 years (the last major earthquake in the region occurred 135 years ago) would lead to a slip of ca. 14 cm (simple calculation for a plunge of 45°). This implicates that a major earthquake along the Polcenigo-Maniago Thrust is possible in this century.

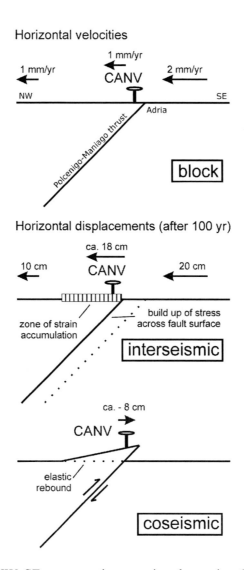

Figure 16. Schematic NW–SE cross-section crossing the station CANV. The velocity and displacement values are estimated and with respect to northern Friuli, and the sketch is not true to scale. CANV is located close to the active Polcenigo-Maniago Thrust (PMT), north of it. Its block velocity should be significantly lower than the one of the Adria (here: 1 mm/a lower). The interseismic horizontal displacement of CANV (18 cm) is affected by the locked PMT and a zone of strain accumulation north of it (shaded). The accumulated strain is released during the coseismic 'backward' jump. The sum of interseismic and co-seismic displacement corresponds the displacement of the block [Bechtold et al. submitted].

5. Conclusion

The processing of 4.5 years of continuous GPS data acquired by FReDNet and several other networks constrain the shortening taking place in the Friuli area. The Friuli GPS velocity field obtained from GPS measurements defines an area that is characterized by 1.6 to 2.2 mm/a of shortening. Furthermore, it indicates oblique collision, with southern Friuli moving NNW (at 330°) towards northern Friuli. The shortening rate and movement direction of southern Friuli/northern Adria is consistent with geomorphologic studies along the Montello Thrust, as well as former-published GPS velocity fields. But since our analysis spans longer time series and includes more local GPS sites (first complete processing of FReDNet), it presents more detail in the region of interest.

Although the seismic activity is currently rather low in the western sector, strain accumulation suggests a higher probability of seismic activity in the future. One should consider that the typical recurrence intervals of earthquakes $M > 6$ in the Friuli area are one order of magnitude longer than the 30 years-long history of seismic monitoring [Galadini et al. 2005]. In the western sector station CANV presents a GPS velocity that is higher than the other stations north of the active thrust front. We propose that the motion of Adria is rigidly transferred via the currently locked Polcenigo-Maniago Thrust (PMT) to the hanging-wall and strain accumulates north of CANV. The coseismic movement of CANV will be SE-directed leading to an overall velocity similar to the long-term geologic movement of northern Friuli, which is now represented by the other northern stations that are not or only slightly affected by locked faults. We recommend a field campaign GPS survey to cross the PMT.

Acknowledgments

B. King (MIT) and T. A. Herring (MIT) provided a number of helpful suggestions for the GPS data processing.

References

[1] Altamimi, Z., Sillard, P., Boucher, C. (2002). ITRF2000: A new release of the International Terrestrial Reference Frame for earth science applications, *J. Geophys. Res.*, **107**(B10), 2214.

[2] Anderson, H. A., Jacksonm, J. A. (1987). Active tectonics of the Adriatic region, *Geophys. J. R. Astron. Soc.*, **91**, 937–983.

[3] Argnani, A., (2006). Some issues regarding the central Mediterranean neotectonics, *Bolletino di Geofisica Teorica ed Applicata*, **47**, 1-2, 13-37.

[4] Battaglia, M., Zuliani, D., Pascutti, D., Michelini, A., Marson, I., Murray, M.H., Bürgmann, R. (2003). Network Assesses Earthquake Potential in Italy's Southern Alps, *EOS*, **84**, 28, 262-264.

[5] Battaglia, M., Murray, M. H., Serpelloni, E., Bürgmann, R. (2004). The Adriatic region: An independent microplate within the Africa-Eurasia collision zone, *Geophys. Res. Lett.*, **31**, L09605.

[6] Bechtold, M., Battaglia, M., Tanner, D. C., Zuliani, D., (submitted). Constraints on the active tectonics of the Friuli/NW-Slovenia area from CGPS measurements and three-dimensional kinematic modeling, *Journal of Geophysical Research, Solid Earth*.

[7] Bennett, R.A., Wernicke, B.P., Davis, J.L., Elosegui, P., Snow, J.K., Abolins, M.J., House, M.A., Stirewalt, G.L., Ferrill, D.A. (1997). Global Positioning System constraints on fault slip rates in the Death Valley region, California and Nevada, *Geophysical Research Letters*, **24**, 23, 3073-3076.

[8] Beutler, G., Mueller, I.I., Neilan, R.E. (1994). The International GPS Service for Geodynamics: development and start of official service on January 1, 1994, *Bulletin Geodesique*, **68**, 39-70.

[9] Blewitt, G., & Lavallée, D. (2002). Effect of annual signals on geodetic velocity, *J. Geophys. Res.*, **107**(B7), 2145.

[10] Bressan, G., Snidarcig, A., Venturini, C. (1998). Present state of tectonic stress of the Friuli area (eastern Southern Alps), *Tectonophysics*, **292**.

[11] Bressan, G., Bragato, P. L., Venturini, C. (2003). Stress and Strain Tensors Based on Focal Mechanisms in the Seismotectonic Framework of the Friuli-Venezia Giulia Region (Northeastern Italy), *Bull. Seismol. Soc. Am.*, **93**, 3.

[12] Calais, E., Nocquet, J.M., Jouanne, F., Tardy, M. (2002). Current strain regime in the western Alps from continuous global positioning system measurements, 1996-2001, *Geology*, **30**, 651–654.

[13] Calais, E., DeMets, C., Nocquet, J.M. (2003). Evidence for a post-3.16 Ma change in Nubia-Eurasia-North America plate motions?, *Earth and Planetary Science Letters*, **216**, 8–92.

[14] Castellarin, A., Nicolich, R., Fantoni, R., Cantelli, L., Sella, M., Selli, L. (2006). Structure of the lithosphere beneath the Eastern Alps (southern sector of the TRANSALP transect), *Tectonophysics*, **414**.

[15] D'Agostino, N., Cheloni, D, Mantenuto, S., Selvaggi, G., Michelini, A., Zuliani, D. (2005). Strain accumulation in the southern Alps (NE Italy) and deformation at the northeastern boundary of Adria observed by CGPS measurements, *Geophys. Res. Lett.*, **32**, L19306.

[16] DeMets, C., Gordon, R. G., Argus, D. F., Stein, S. (1994). Effect of recent revisions to the geomagnetic reversal time scale on estimates of current plate motions, *Geophysical Research Letters*, **21**, 20, 2191–2194.

[17] Dewey, J.F., Pitman W.C., Ryan, W.B.F., Bonnin, J. (1973). Plate tectonics and the evolution of the Alpine system, *Geological Society of America Bulletin*, **84**, 3137–3180.

[18] Dewey, J.F., Helman, M.L., Turco, E., Hutton, D.H.W., Knott, S.D. (1989). Kinematics of the Western Mediterranean, in Alpine Tectonics, *Geological Society of London Special Publications*, Coward, M.P., Dietrich, D., Park, R.G. (eds), 45, 265–283.

[19] DISS Working Group (2006). Database of Individual Seismogenic Sources (DISS), version 3: A compilation of potential sources for earthquakes larger than M 5.5 in Italy and surrounding areas, © INGV 2005, 2006 - Istituto Nazionale di Geofisica e Vulcanologia - All rights reserved. http://www.ingv.it/DISS/.

[20] Dixon, T.H., Mao, A. (1997), A GPS estimate of relative motion between North and South America, *Geophysical Research Letters*, **24**, 535–538.

[21] Dong, D., Herring, T.A., King, R.W. (1998), Estimating regional deformation from a combination of space and terrestrial geodetic data, *Journal of Geodesy*, 72, 200–214.

[22] Feigl, K. L., et al. (1993), Space geodetic measurement of crustal deformation in central and southern California, 19841992, *J. Geophys. Res.*, **98**(B12), 21, 67721, 712.

[23] Gahalaut, V.K., Chander, R. (1997), On interseismic elevation changes for great thrust earthquakes in the Nepal Himalaya, *Geophysical Research Letters*, **24**, 9, 1011–1014.

[24] Galadini, F., M. E. Poli, A. Zanferrari (2005), Seismogenic sources potentially responsible for earthquakes with M > 6 in the eastern Southern Alps (Thiene-Udine sector, NE Italy), *Geophys. J. Int.*, **161**(3).

[25] Gatti, M., Stoppini, A. (2000), Appropriate use of international reference frames in regional GPS applications: guidelines and examples, *Bollettino di Geodesia e Scienze Affini*, **1**, 1–19.

[26] Herring, T. A. (2003a), GLOBK: Global Kalman filter VLBI and GPS analysis program, version 10.1, 91 pp., Massachusetts Institute of Technology.

[27] Herring, T. A. (2003b), MATLAB Tools for viewing GPS velocities and time series, *GPS Solutions*, **7**(3), 194–199.

[28] Herring, T.A., Davis, J.L., Shapiro, I.I. (1990), Geodesy by Radio Interferometry: The Application of Kalman Filtering to the Analysis of Very Long Baseline Interferometry data, *Journal of Geophysical Research*, **95**, B8, 561–581.

[29] Johnson, H.O., Agnew, D.C. (1995), Monument motion and measurements of crustal velocities, *Geophysical Research Letters*, **22**, 2905–2908.

[30] Langbein, J. (2004), Noise in two-color electronic distance meter measurements revisited, *Journal of Geophysical Research*, **109**, B4, pp. 16.

[31] Larson, K.M., Freymueller, J.T., Philipsen, S. (1997), Global plate velocities from the Global Positioning System, *Journal of Geophysical Research*, **102**, 9961–9981.

[32] Le Pichon, X., Mazzotti, S., Henry, P., Hashimoto, M. (1998), Deformation of the Japanese Islands and seismic coupling: an interpretation based on GSI permanent GPS observations, *Journal of Geophysical Research*, **134**, 2, 501–514.

[33] Le Pichon, X., Kreemer, C. (2004), The effects of the definition of a stable Eurasian reference frame on geodetic velocity estimates along the its boundaries, *AGU/CGU Joint Assembly, Abstract G24A-05*.

[34] Mao, A., Harrison, C.G.A., Dixon, T.H. (1999), Noise in GPS coordinate time series, *Journal of Geophysical Research*, **104**, 2797–2816.

[35] McClusky, S., et al. (2000), Global Positioning System constraints on plate kinematics and dynamics in the eastern Mediterranean and Caucasus, *J. Geophys. Res.*, **105**(B3), 5695-5720.

[36] Meade, B.J., Hager, B.H. (2005), Block models of crustal motion in southern California constrained by GPS measurements, *Journal of Geophysical Research*, **110**, B3, pp. 19.

[37] Miller, M.M., Johnson, D.J., Rubin, C.M., Dragert, H., Wang, K., Qamar, A., Goldfinger, C. (2001), GPS-determination of along-strike variation in Cascadia margin kinematics: Implications for relative plate motion, subduction zone coupling, and permanent deformation, *Tectonics*, **20**, 2, 161–176.

[38] Niell, A.E. (1996), Global mapping functions for the atmospheric delay at radio wavelengths, *Journal of Geophysical Research*, **101**, 3227–3246.

[39] Nikolaidis, R. (2002), Observation of Geodetic and Seismic Deformation with the Global Positioning System, *Ph.D. Thesis, University of California, San Diego,* pp. 249.

[40] Nocquet, J.M., Calais, E. (2004), Geodetic Measurements of Crustal Deformation in the Western Mediterranean and Europe, *Pure and Applied Geophysics*, **161**, 3, 861–880.

[41] Oldow, J.S., Ferranti, L., Lewis, D.S., Campbell, J.K., D'Argenio, B., Catalano, R., Pappone, G., Carmignani, L., Conti, P., Aiken, C.L.V. (2002), Active fragmentation of Adria, the north Africa promontory, central Mediterranean orogen, *Geology*, **30**, 9, 779–782.

[42] Oral, M.B. (1994), Global Positioning System (GPS) Measurements in Turkey (1988-1992): Kinematics of the Africa-Arabia-Eurasia Plate Collision Zone, *Ph.D. Thesis, Massachusetts Institute of Technology, USA,* pp. 344.

[43] Poli, M. E., L. Peruzza, A. Rebez, G. Renner, D. Slejko, A. Zanferrari (2002), New seismotectonic evidence from the analysis of the 1976-1977 and 1977-1999 seismicity in Friuli (NE Italy), *Bolletino di Geofisica Teorica ed Applicata*, **43**, 53–78.

[44] Poli, M. E. and G. Renner (2004), Normal focal mechanisms in the Julian Alps and Prealps: seismotectonic implications for the Italian-Slovenian border region, *Bollettino di Geofisica Teorica ed Applicata*, **45**(1-2), 51–69.

[45] Reilinger, R., et al. (2006), GPS constraints on continental deformation in the Africa-Arabia-Eurasia continental collision zone and implications for the dynamics of plate interactions, *J. Geophys. Res.*, **111**.

[46] Roeder, D., Lindsey, D. (1992), Barcis area (Veneto, Friuli, Slovenia): architecture and geodynamics, *Nafta*, **43**, 509–548.

[47] Savage, J.C., Burford, R.O. (1973), Geodetic determination of the relative plate motion in central California, *Journal of Geophysical Research*, **78**, 832–845.

[48] Serpelloni, E., Anzidei, M., Baldi, P., Casula, G., Galvani, A., Pesci, A., Riguzzi, F. (2002), Combination of permanent and non-permanent GPS networks for the evaluation of the strain-rate field in the central Mediterranean area, *Bolletino di Geofisica Teorica ed Applicata*, **43**, 3-4, 195–219.

[49] Serpelloni, E., Anzidei, M., Baldi, P., Casula, G., Galvani, A. (2005), Crustal velocity and strain-rate fields in Italy and surrounding regions: new results from the analysis of permanent and non-permanent GPS networks, *Geophysical Journal International*, **161**, 3, 861–880.

[50] Serpelloni, E., Vannucci, G., Pondrelli, S., Argnani, A., Casula, G., Anzidei, M., Baldi, P., Gasperini, P. (2007), Kinematics of the Western Africa-Eurasia plate boundary from focal mechanisms and GPS data, *Geophys. J. Int.*, **169**(3), 1180–1200.

[51] Slejko, D., G. Neri, I. Orozova, G. Renner, M. Wyss (1999), Stress field in Friuli (NE Italy) from fault plane solutions of activity following the 1976 main shock, *Bull. Seismol. Soc. Am.*, **89**(4), 1037–1052.

[52] Tregoning, P., van Dam, T. (2005), Atmospheric pressure loading corrections applied to GPS data at the observation level, *Geophysical Research Letters*, 32, 1–4.

[53] Turcotte, D.L., Schubert, G. (2002), Geodynamics: Application of Continuum Physics to Geological Problems, *2nd ed, Cambridge University Press,* pp. 456.

[54] Vannucci, G., Pondrelli, S., Argnani, A., Morelli, A., Gasperini, P., Boschi, E. (2004) An atlas of Mediterranean seismicity, *Annals of Geophysics, Supplement to Vol. 47, N. 1,* 247–306.

[55] Working Group CPTI (1999), Catalogo parametrico dei terremoti italiani, *ING, GNDT, SGA, SSN, Bologna,* pp. 92.

[56] Zhang, J., Bock, Y., Johnson, H., Fang, P., Williams, S., Genrich, J., Wdowinski, S., Bahr, J. (1997), Southern California permanent GPS geodetic array: Error analysis of daily position estimates and site velocities, *Journal of Geophysical Research*, **102**, 18035–18055.

In: Global Positioning Systems
ISBN 978-1-60741-012-6
Editors: V. Asphaug and E. Sørensen, pp. 157-170 © 2010 Nova Science Publishers, Inc.

Chapter 7

ESTIMATION OF REGIONAL STRESS INCREMENT DISTRIBUTION USING GPS ARRAY DATA

Muneo Hori[1], Takeshi Iinuma[2] and Teruyuki Kato[1]
[1]Earthquake Research Institute, University of Tokyo
(Yayoi, Bunkyo, Tokyo 113-0032, Japan)
[2]Research Center for Prediction of Earthquakes and Volcanic Eruptions,
Graduate School of Science, Tohoku University

Abstract

This article presents an analysis method of GPS array data which is aimed at estimating stress increment filed that is associated with displacement or strain increment field that is observed by the GPS array. The method is based on an inverse analysis which finds self-equilibrating stress for a body with unknown stress-stress relation, when a distribution of strain is measured. With the assumptions of the plane stress state and no volumetric inelastic deformation for the incremental deformation of the Japanese Islands, the stress increment is computed from data which are measured by the GPS array that has been operating in Japan. It is shown that the stress increment is not uniform and that there are some regions which have sharp changes in the stress increment; these changes are associated with changes in observed strain increment even though the distribution of the estimated stress increment is different from that of the strain increment which is computed from the GPS array data. Some discussions are made for the usefulness and limitations of applying the present data analysis to the GPS array data.

1. Introduction

Advancement of computational mechanics enables us to solve a solid continuum problem by means of numerical computation even for a body with complicated configuration or subjected to various loadings. A remaining issue in computational mechanics is better understanding of material properties. A new material test is needed for such understanding; a target of the material test is an extremely small object such as new electronic devices or human organisms. Another kind of the target is the Earth. The deformation of the Earth, which leads to local failure or earthquakes, is reproduced or predicted with higher accuracy as the Earth's material properties are identified more accurately and precisely.

Global Positioning System (GPS) could be used as a device of a new material test of the Earth. While the final goal is to identify local constitutive relation or stress-strain relation, as the first step, we seek to identify the local stress distribution by analyzing data obtained by an array of GPS stations, which have been installed in Japan to monitor the crustal deformation of the Japanese islands[5]. The array measures regional displacement or strain increment, but not cannot estimate stress increment that is associated with strain increment since regional stress-strain relations are not known.

When the stress-strain relation is not known, the identification of stress using strain is regarded as an ill-posed inverse problem; see Bui's work[2]. It is true that stress is determined from strain if the stress-strain relation is known. However, this does not imply that stress cannot be determined from strain if the stress-strain relation is not known. We thus consider a mathematical problem, which has a unique solution, to determine stress from strain even if the stress-strain relation is only partially known. More specifically, we consider the following problem:

> For a body in a state of plane stress, identify the distribution of stress that satisfies the equation of equilibrium, using the distribution of strain and one stress-strain relation.

To solve this problem, we have to take full advantage of the equilibrium equation that stress satisfies. The number of the stress components and the equilibrium equation in plane stress state is three and two, respectively. Thus, one equation, which is independent from the equilibrium equation, is needed to determine the three stress components. One relation between stress and strain (or partial information of the stress-strain relation) serves as this equation.

As mentioned, the above problem is the identification of stress from strain using partial information of stress-strain relation, and could be regarded as an inverse problem; see a list of related references[1, 9, 12]; see also references[3, 10, 11]. Thus, we name a method of solving the above problem as a stress inversion method[4, 6].

The stress inversion method presented here is applicable only to the two-dimensional setting. Extension to the three-dimensional setting will be difficult unless three-dimensional distribution of strain is measured. We should mention that the plane stress state is an approximate state that is applicable to a thin body. Thus, we analyze stress within a shallow part of the crust. The size of the Japanese Island is of the order of 100[km], and hence the thickness of the shallow part will be a few kilo meter. And what is analyzed by the stress inversion method is stress that is averaged in the vertical direction. Also, we should mention that the GPS array monitors the increment of the displacement, and hence the stress inversion method computes the stress increment that is associated with the observed displacement increment or strain increment.

This article presents the formulation of the stress inversion method. Several examples of applying this method to the GPS array data are presented as well. The content of the article is as follows: In Section 2, the stress inversion method is formulated; instead of using stress tensor components, Airy's stress function which produces self-equilibrating stress components for a body in a state of plane stress, is used. In Section 3, explained are some assumptions which are made in applying the stress inversion method to the GPS array data. Results of numerical analysis of the GPS array data by means of the stress inversion method are presented, too. Concluding remarks are made in Section 4.

This paper uses two-dimensional Cartesian coordinates, denoted by x_i. Index notation is used for a vector or tensor quantity, the summation convention is employed, and indices following a comma denote partial differentiation with respect to the corresponding coordinates.

2. Formulatoin of Stress Inversion Method

We consider a thin body which is in a state of plane stress, and denote the surface and the boundary of the body by S and ∂S, respectively; see Fig. 1. While the stress-strain relation is not known for this body, it is assumed that one relation between the two-dimensional dilatant or volumetric stress and strain, $\sigma = \sigma_{ii}$ and $\epsilon = \epsilon_{ii}$, is known. For simplicity, the relation is linear, i.e.,

$$\sigma = \kappa \epsilon, \tag{1}$$

where κ is the two-dimensional bulk moduli and uniform in S.

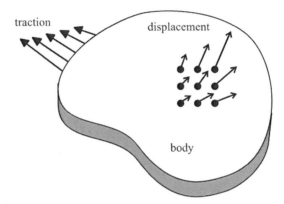

Figure 1. Body in plane stress state and data input to stress inversion method.

The equation of equilibrium for the three stress components is automatically satisfied if the components are expressed in terms of Airy's stress function[1]. Instead of seeking the three stress components, we seek to determine this single function. Denoting this function by a, we can write the stress components as

$$\begin{bmatrix} \sigma_{11} \\ \sigma_{22} \\ \sigma_{12} \end{bmatrix} = \begin{bmatrix} a_{,22} \\ a_{,11} \\ -a_{,12} \end{bmatrix}. \tag{2}$$

The assumed stress-strain relation, Eq. (1), is used as a governing equation for a. Indeed, we can derive the governing equation of a as

$$a_{,ii}(\mathbf{x}) = \kappa \epsilon(\mathbf{x}) \quad \text{for } \mathbf{x} \text{ in } S. \tag{3}$$

---

[1] In general, a governing equation of Airy's stress function is derived from the compatibility condition of strain when the stress-strain relation is fully known. For instance, when the stress-strain relation is linearly isotropic, the governing equation of Airy's stress function becomes a homogeneous bi-harmonic equation.

Note that the right side of the equation is given when local strain is measured. It will be transparent that another governing equation of a is derived if another stress-strain relation is known. A possible candidate of such stress-strain relation is isotropy; isotropy implies that the direction of the maximum shear stress is parallel to that of the maximum shear strain.

The governing equation of a, Eq. (3),is Poisson's equation. If suitable boundary conditions are given on the body's boundary, ∂S, we can uniquely determine a by solving the resulting boundary value problem. For instance, if tractions are measured on ∂S, such boundary conditions are obtained with some manipulation. Denoting the boundary traction and the unit outer normal by t_i and n_i, respectively, we have $t_j = n_i \sigma_{ij}$ or

$$\begin{bmatrix} t_1 \\ t_2 \end{bmatrix} = \begin{bmatrix} n_1 a_{,22} - n_2 a_{,12} \\ -n_1 a_{,12} + n_2 a_{,11} \end{bmatrix}. \tag{4}$$

Denoting the unit tangential vector by s_i and using $[s_1, s_2]^T = [-n_2, n_1]^T$, we rewrite the left side of Eq. (4) as

$$\begin{bmatrix} n_1 a_{,22} - n_2 a_{,12} \\ -n_1 a_{,12} + n_2 a_{,11} \end{bmatrix} = \begin{bmatrix} \frac{\partial a_{,2}}{\partial s} \\ -\frac{\partial a_{,1}}{\partial s} \end{bmatrix}.$$

Thus, integration of the boundary tractions leads to the expression of the resultant force, $r_i = \int t_i \, ds$, in terms of a as

$$\begin{bmatrix} r_1 \\ r_2 \end{bmatrix} = \begin{bmatrix} a_{,2} \\ -a_{,1} \end{bmatrix}.$$

The derivative of a is given by the resultant force, and hence Neumann boundary conditions for a are obtained. That is,

$$n_i(\mathbf{x}) a_{,i}(\mathbf{x}) = -\varepsilon_{3ij} n_i(\mathbf{x}) r_j(\mathbf{x}) \quad \text{for } \mathbf{x} \text{ in } S, \tag{5}$$

where ε_{ijk} is a permutation symbol, i.e., $\varepsilon_{3ij} n_i r_j = n_1 r_2 - n_2 r_1$.

The governing equation and the boundary conditions, Eqs. (3) and (5), form a linear boundary value problem for a. That is,

$$\begin{cases} a_{,ii}(\mathbf{x}) = \kappa \bar{\epsilon}(\mathbf{x}) & \text{for } \mathbf{x} \text{ in } S, \\[2mm] n_i(\mathbf{x}) a_{,i}(\mathbf{x}) = -\varepsilon_{3ij} n_i(\mathbf{x}) \bar{r}_j(\mathbf{x}) & \text{for } \mathbf{x} \text{ on } \partial S. \end{cases} \tag{6}$$

Here, bar on strain and resultant force emphasizes that they are the measured data.

We must pay attention to the *consistency* of the boundary value problem with the Neuman boundary conditions. Since the governing equation is Poisson's equation, the consistency means that the surface integration of $\kappa \bar{\epsilon}$ coincides with the line integration of $-\varepsilon_{3ij} n_i \bar{r}_j$. This is easily proved since

$$\int_S a_{,ii}(\mathbf{x}) \, ds = \int_{\partial S} n_i(\mathbf{x}) a_{,i}(\mathbf{x}) \, d\ell.$$

The consistency implies that the data of strain and resultant force must satisfy

$$\int_S \kappa \bar{\epsilon}(\mathbf{x}) \, ds = \int_{\partial S} -\varepsilon_{3ij} n_i(\mathbf{x}) \bar{r}_j(\mathbf{x}) \, d\ell. \tag{7}$$

Estimation of Regional Stress Increment Distribution Using GPS Array Data   161

If the data that satisfy Eq. (7) are given, Airy's stress function is determined by solving Eq. (6), and hence the stress components are obtained. Due to the nature of the Neumann boundary conditions, an arbitrary constant can be added to the solution of a. The constant, however, does not change the stress components since they are computed by taking the derivative of a; see Eq. (2).

The stress inversion method leads to the boundary value problem for a, Eq. (6), which is linear even if stress-strain relation of the target body is not linear. This inversion method is applicable to any arbitrary body with non-linear stress-stain relation or material properties. The authors have applied this linear boundary value problem to a non-linear elasto-plastic material sample in order to estimate the plastic deformation[4].

It is interesting to extend the stress inversion method to a body at dynamic state. The equation of equilibrium is replaced by the equation of motion, i.e.,

$$\sigma_{ij,i}(\mathbf{x},t) = \rho\,\ddot{u}_i(\mathbf{x},t),$$

where $\ddot{u}_i = \frac{\partial^2 u_i}{\partial t^2}$ is acceleration and ρ is density of the material. When spatial distribution and temporal change of displacement is measured, data of strain and acceleration are given. Thus, the governing equations for the stress components are written as

$$\begin{cases} \sigma_{ij,i}(\mathbf{x},t) = \rho\ddot{\bar{u}}_i(\mathbf{x},t) & (i=1,2), \\ \sigma_{ii}(\mathbf{x},t) = \kappa\,\bar{\epsilon}(\mathbf{x},t), \end{cases} \quad \text{for } \mathbf{x} \text{ in } S \text{ and } t > 0. \tag{8}$$

The boundary conditions are the same as Eq. (5) and suitable initial conditions must be found. Together with these boundary and initial conditions, the three equations of Eq. (8) form a linear initial-boundary value problem for the three stress components. This linear problem is applicable to non-linear materials, just as Eq. (6) is applicable to non-linear materials at quasi-static state.

3. GPS Array Data Analysis

It is not necessary to explain a nation-wide GPS array that consists of more than 1,000 stations and has been operating in Japan since 1994; see [5]; see Fig. 2. The displacement distribution itself provides vital data for the crustal deformation. With new data analysis methods, further information might be obtained for the mechanism of the crustal deformation that triggers earthquakes. The authors are applying the stress inversion method to the GPS array data, to develop a system which monitors the change in regional strain as well as regional stress that is associated with the regional strain.

3.1. Assumptions Needed in Applying Stress Inversion Method to GPS Array Data

The stress inversion method is not a tool that automatically computes stress from measured strain data. The method has some limitations. First of all, the assumption of plane stress state is necessary. Thus, in applying the stress inversion method to the GPS array data, we focus a shallow part of the crust so that stress which is averaged in the vertical direction is a target of the analysis. Even if a shallow part of the crust is considered, the assumption of

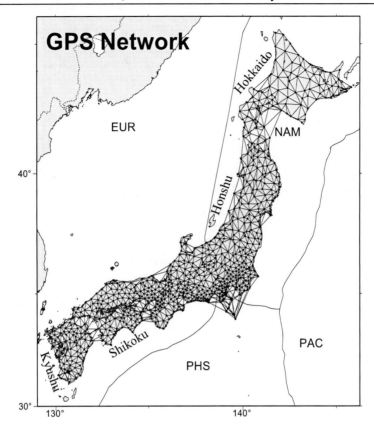

Figure 2. GPS array of the Japanese Islands.

plane stress state is *strong* since the islands are a part of Eurasia Plate under which Pacific and Philippine Sea Plates are subducting; North America Plate is pushing the Eurasia Plate; and the curvature of the Earth needs to be taken into consideration. However, making this assumption could be accepted since the surface of the Japanese Islands remain traction free and the effects of gravity and past plate movement on the islands' *permanent* deformation are not observed by the GPS array; in the time scale of the GPS array operation, deformation due to gravity or deformation caused by the past plate movement is regarded as permanent.

In applying the stress inversion method to the GPS array data, we use increment formulation, in order to emphasize the fact that the displacement measured by the array is the displacement increment during the observation period; the array cannot measure permanent displacement. Since the array provides displacement increment at the GPS stations, a strain increment field can be computed by interpolating the measured displacement increment and taking spatial differentiation.

There are two other tasks needed to apply the stress inversion method to the GPS array data. The first task is to determine the constant that relates the dilatant stress and strain increment. Note that setting a linear (elastic) relation between dilatant shear and stress is based on the assumption that the deformation of the Japanese Islands is elastic except for shear deformation that is caused by sliding of faults. Thus, while linearity is not assumed

Estimation of Regional Stress Increment Distribution Using GPS Array Data 163

for shear deformation, we can use a linear relation for dilatant shear and stress, i.e.,

$$d\sigma = \kappa \, d\epsilon. \tag{9}$$

This assumption is applicable to incompressible elasto-plastic materials if $d\epsilon_{33}$ is neglected. This vertical strain component is fully neglected in this article, but we can measure this component by using another space monitoring technique such as SAR; see [8].

The second task is to prescribe the boundary conditions. Since measuring the boundary traction of the Japanese Islands is impossible, we make another assumption that the boundary traction rate is generated by a certain (unknown) uniform stress increment. This assumption is regarded as the first-order approximation which uses the average of the stress increment; actually, the movement of the four plates near the Japanese Islands produces complicated distribution of stress increment on the boundary. To derive boundary conditions, we introduce Airy's stress function which generates a uniform stress increment field, i.e.,

$$da^\circ(\mathbf{x}) = \frac{1}{2}(x_2^2 \, d\sigma_{11}^0 - 2x_1 x_2 \, d\sigma_{12}^0 + x_2^2 \, d\sigma_{22}^0),$$

where $d\sigma_{ij}^0$ is unknonw stress increment. Then, the boundary conditions are written as

$$n_i(\mathbf{x}) \, da_{,i}(\mathbf{x}) = n_i(\mathbf{x}) \, da_{,i}^\circ(\mathbf{x})$$

on ∂S. The unknown stress increment is determined by using the consistency condition, Eq. (7). Indeed, Eq. (7) leads to

$$\int_S \kappa \, d\bar{\epsilon}(\mathbf{x}) \, ds = S(d\sigma_{11}^0 + d\sigma_{22}^0),$$

where S stands for the area of the surface S. If the form of $d\sigma_{ij}^0$ is assumed as $d\sigma_{ij}^0 = d\sigma^\circ \delta_{ij}$, this $d\sigma^\circ$ is determined as

$$d\sigma^\circ = \frac{\kappa}{2S} \int_S d\bar{\epsilon}(\mathbf{x}) \, ds. \tag{10}$$

Hence, the boundary conditions are finally expressed in terms of this $d\sigma^\circ$ as

$$n_i(\mathbf{x}) da_{,i}(\mathbf{x}) = (n_1(\mathbf{x})x_2 + n_2(\mathbf{x})x_1) \, d\sigma^\circ. \tag{11}$$

The boundary value problem for an increment of Airy's stress function is thus constructed in view of Eqs. (9) and (11), as follows:

$$\begin{cases} da_{,ii}(\mathbf{x}) = \kappa \, d\bar{\epsilon}_{ii}(\mathbf{x}) & \text{for } \mathbf{x} \text{ in } S, \\ n_i(\mathbf{x}) \, da_{,i}(\mathbf{x}) = (n_1(\mathbf{x})x_2 + n_2(\mathbf{x})x_1) \, d\sigma^\circ & \text{for } \mathbf{x} \text{ on } \partial S. \end{cases} \tag{12}$$

Here, $d\sigma^\circ$ is explicitly computed when $d\epsilon$ is given; see Eq. (10). In order to solve this boundary value problem numerically, we consider a weak form of this problem, using a suitable weight function, ϕ. The weak form is $\int_S \phi(da_{,ii} - \kappa d\bar{u}_{i,i}) \, ds = 0$, or

$$\int_S \phi_{,i}(\mathbf{x})(da_{,i}(\mathbf{x}) - \kappa d\bar{u}_i(\mathbf{x})) \, ds +$$

$$\int_{\partial S} \phi(\mathbf{x})((n_1(\mathbf{x})x_2 + n_2(\mathbf{x})x_1)d\sigma^\circ - \kappa n_i(\mathbf{x})d\bar{u}_i(\mathbf{x})) \, d\ell = 0. \tag{13}$$

As is seen, the displacement increment that are measured by the GPS array are used in Eq. (13). The strain increment does not appear in this weak form and hence it is not necessary to compute spatial derivative of displacement increment to obtain strain increment.

Once da is obtained from Eq. (13), it is straightforward to compute the regional strain and stress increment. For a triangle region Ω which is formed by connecting three neighboring GPS stations, the average of regional strain and stress increment are

$$
\begin{bmatrix} \langle d\epsilon_{11} \rangle \\ \langle d\epsilon_{22} \rangle \\ \langle d\epsilon_{12} \rangle \end{bmatrix} = \frac{1}{\Omega} \int_{\partial\Omega} \begin{bmatrix} n_1(\mathbf{x})d\overline{u}_1(\mathbf{x}) \\ n_2(\mathbf{x})d\overline{u}_2(\mathbf{x}) \\ \frac{1}{2}(n_1(\mathbf{x})d\overline{u}_2(\mathbf{x}) + n_2(\mathbf{x})d\overline{u}_1(\mathbf{x})) \end{bmatrix} d\ell
$$

and

$$
\begin{bmatrix} \langle d\sigma_{11} \rangle \\ \langle d\sigma_{22} \rangle \\ \langle d\sigma_{12} \rangle \end{bmatrix} = \frac{1}{\Omega} \int_{\partial\Omega} \begin{bmatrix} n_2(\mathbf{x})da_{,2}(\mathbf{x}) \\ n_1(\mathbf{x})da_{,1}(\mathbf{x}) \\ -n_2(\mathbf{x})da_{,1}(\mathbf{x}) \end{bmatrix} d\ell.
$$

As is seen, these average quantities are computed by using the displacement increment and the gradient of Airy's stress increment, without taking higher derivatives.

3.2. Results of Numerical Analysis of GPS Array Data

The stress inversion method is now applied to the GPS array data. A finite element method with a triangular element of 15 nodes is used to numerically solve Eq. (13); see [8]. A triangular element is formed by connecting three neighboring GPS network stations, and other twelve nodes are equally spaced in the element. Meshing is made so that triangles become as close to be equilateral as possible. The displacement increment measured at each station is assigned to the corner points, and other nodes have linearly interpolated values; see Fig. 3 for the displacement increment distribution which is used in the present analysis.

First, the distribution of regional average strain increment is plotted in Fig. 4; a) and b) are for the dilate strain increment and for the maximum shear strain increment,

$$
d\gamma = \sqrt{\frac{1}{4}(d\overline{\epsilon}_{11} - d\overline{\epsilon}_{22})^2 + d\overline{\epsilon}_{12}^2},
$$

respectively. The distribution of strain increment is not uniform. In particular, it is shown that there are some regions where the strain increment is concentrated. Such large spatial change in strain increment has not been observed. The accuracy of the strain increment distribution that is computed from the GPS array data should be examined by comparing with other geodesic data, such as the measurement of strain meters. Rigid body motion of neighboring GPS stations are removed by computing the average strain increment, and the strain increment distribution shown in Fig. 4 is theoretically free from noises which correspond to such rigid body motion. A further study is needed to get rid of noises in strain increment distribution which come form other sources.

Next, the distribution of regional average stress increment is plotted in Fig. 5; a) and b) are for the volumetric stress increment and for the maximum shear stress increment,

$$
d\tau = \sqrt{\frac{1}{4}(d\sigma_{11} - d\sigma_{22})^2 + d\sigma_{12}^2}.
$$

Estimation of Regional Stress Increment Distribution Using GPS Array Data   165

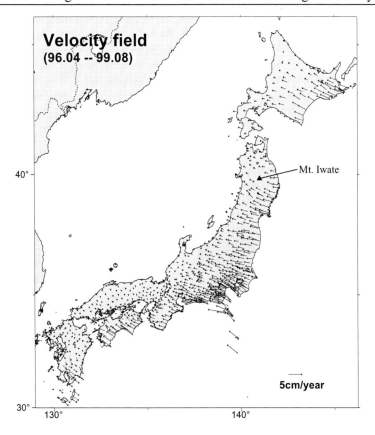

Figure 3. Displacement increment measured by GPS array.

The value of κ is set as $\kappa = 200$[GPa]. This stress increment distribution is actually computed using displacement increment, even though it is associated with the strain increment distribution. Comparing Fig. 5 with Fig. 5, it is seen that the stress increment distribution is slightly different from the strain increment distribution, even though overall patterns are similar to each other. If *apparent* stress increment is computed by simply multiplying the measured strain increment with a certain elasticity tensor, it does not satisfy the equation of equilibrium. The stress increment obtained by the stress inversion method satisfies the equation of equilibrium. Therefore, the stress and strain distributions are slightly different from each other. It is certainly true that examining[2] the accuracy of the computed stress increment distribution is difficult unless regional average stress increment is actually measured and the measured stress increment is compared with the computed one. However, it might be worth monitoring the stress increment that is computed by applying the stress inversion method to the GPS array data, in order to draw some information about the stress increment change that is associated with the observed crustal deformation. The crustal deformation monitor that is developed by the authors continuously checks the change in the

---

[2] As summarized in the preceding subsection, several assumptions have been made in applying the stress inversion method to the GPS array data. Thus, verifying the stress increment distribution actually means the examinations of the assumptions made.

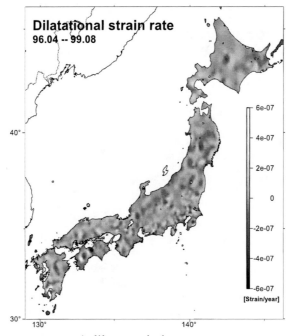

a) dilate strain increment

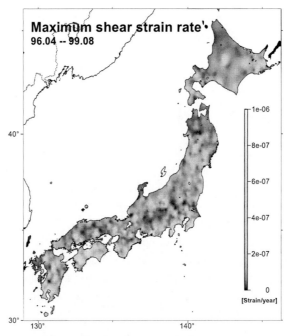

b) maximum shear strain increment

Figure 4. Distribution of strain increment.

Estimation of Regional Stress Increment Distribution Using GPS Array Data   167

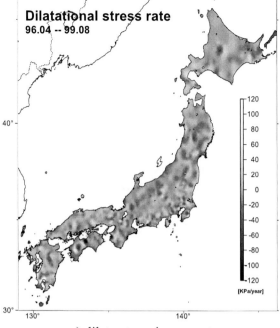

a) dilate stress increment

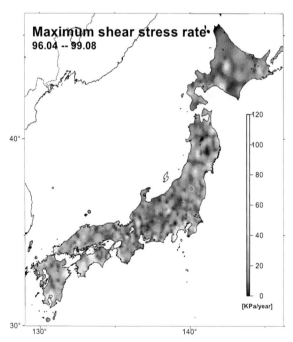

b) maximum shear stress increment

Figure 5. Distribution of stress increment.

average strain and stress increment.

Wild spatial change in the strain increment distribution is observed when the regionally averaged strain increment is computed by using the line integration of the measured displacement along the triangle element edges. The stress increment distribution obtained by the stress inversion method also shows similar regional wild changes although the distribution is different from place to place. In the present analysis, we assume that the two-dimensional bulk modulus, κ, is uniform over the Japanese Islands. This assumption might be relaxed, and we may set a more complicated (and more realistic) condition for κ. Data analysis with the aid of the stress inversion method with a more sophisticated setting will contribute to construct a more rational model of the Japanese Islands. As a trial, we assume the isotropy of the Japanese Islands, abandoning the assumption that κ is uniform. As mentioned, the isotropy implies the coaxiality of the computed stress increment tensor with the measured strain increment tensor, i.e.,

$$\frac{da_{,22}(\mathbf{x}) - da_{,11}(\mathbf{x})}{da_{,12}(\mathbf{x})} = -\frac{d\bar{\epsilon}_{11}(\mathbf{x}) - d\bar{\epsilon}_{22}(\mathbf{x})}{d\bar{\epsilon}_{12}(\mathbf{x})}. \tag{14}$$

Using this equation as the governing equation, we solve the boundary value problem of da and obtain the stress increment distribution. The two dimensional bulk moduli, which is not uniform, is computed as

$$\kappa(\mathbf{x}) = \frac{d\sigma_{ii}(\mathbf{x})}{d\bar{\epsilon}_{ii}(\mathbf{x})}. \tag{15}$$

The distribution of this κ is plotted in Fig. 6. The validity of assuming the isotropy should be discussed, as well as the validity of assuming Eq. (1). At this moment, this figure shows a non-uniform distribution of κ. Since κ is related to the P-wave velocity, it is worth comparing this distribution with the velocity structure of the Japanese Islands which is constructed by analyzing earthquake waves in order to examine the validity of the isotropy assumption.

4. Concluding Remarks

This article presents the stress inversion method as an analysis method of the GPS array data, in order to estimate a distribution of stress increment which is associated with observed displacement increment or strain increment. Some assumptions must be made in applying the stress inversion method, and the validity of the assumptions and hence the results of the GPS array data analysis by means of the stress inversion method are not verified at this moment. However, new information about the crust deformation is obtained, for self-equilibrating stress increment, the spatial concentration of which is different from that of the associated strain increment. Thus, the method serves as a data analysis method to monitor the change in crustal deformation of the Japanese Islands.

In closing this article, we emphasize that the stress inversion method solves a mathematically well-posed problem and that there is no ambiguity in solving the problem when sufficient sets of data are provided. From the geodesy view-point, however, the assumptions that are made in applying this method to the GPS array data must be carefully examined. It would be interesting to investigate the validity of these assumptions and the results obtained by means of the stress inversion analysis, by comparing the temporal change in stress

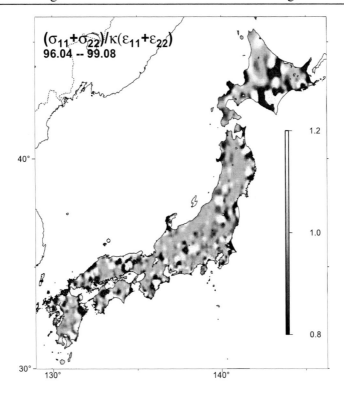

Figure 6. Distribution of computed κ.

increment that is obtained from the GPS array data analysis with curst stress which is continuously measured at several sites.

Acknowedgement

This research is supported partially by Grant-in-Aid for Scientific Research, the Ministry of Education, Science, Sports and Culture and partially by Japan Science and Technology Corporation.

The GPS network data are provided by the Geological Survey Institute.

References

[1] Anikonov, Y. E. 1995. *Multidimensional Inverse and Ill-posed Problems for Differential Equations*, VSP International Science Publishers, New York.

[2] Bui, H. D. 1994. *Inverse Problem in the Mechanics of Materials: An Introduction*, CRC Press, New York.

[3] Eskin, G. and Ralston, J., 2002. On the inverse boundary value problem for linear isotropic elasticity, *Inverse Problems*, **18**, 907-921.

[4] Hori, M. and Kameda, T., 2001, Inversion of stress from strain without full knowledge of constitutive relations, *J. Mech. Phys. Solids*, **49**, 1621-1638.

[5] Geophysical Survey Institute, the Ministry of Construction, Japan, 2000. Home Page, http://mekira.gsi-mc.go.jp/.

[6] Hori, M., Kameda, T. and Kato, T., 2001. Application of stress inversion method to predict stress distribution in Japanese Islands, *Geophys. J. Int.*, **144**, 597-608.

[7] Hori, M., 2003. Inverse analysis method to find local constitutive relations, *Mechanics of Materials*, **35**, 1089-1105.

[8] Iinuma, T. Kato, T. and Hori, M., 2005. Inversion of GPS velocity and seismicity data to yield changes in stress in the Japanese Islands, *Geophys. J. Int.*, **160**, 417-434, 2005.

[9] Kirsch, A., 1996. *An Introduction to the Mathematical Theory of Inverse Problems*, Springer, New York.

[10] Nakamura, G. and Uhlmann, G., 1994, Global uniqueness for an inverse boundary problem arising in elasticity, *Invent. Math.*, **118**, 457-474.

[11] Nakamura, G. and Uhlmann, G., 1995, Inverse problems at the boundary for an elastic medium, *SIAM J. Math. Anal*, **26**, 263-279.

[12] Tanaka, M. and Dulikravich, G.S. (eds.) 1998, *Inverse Problems in Engineering Mechanics*, Elsevier, New York.

In: Global Positioning Systems
Editors: V. Asphaug and E. Sørensen, pp. 171-238 © 2010 Nova Science Publishers, Inc.

ISBN 978-1-60741-012-6

Chapter 8

GPS-BASED OPTIMAL FIR FILTERING AND STEERING OF CLOCK ERRORS

Yuriy Shmaliy
Fimee, Guanajuato Univ., Salamanco Gto., Mexico

Abstract

This Chapter addresses novel results in optimal finite impulse response (FIR) filtering, prediction, and steering of the local clock time interval error (TIE) using the Global Positioning System (GPS) one pulse per second (1PPS) timing signals. Studies are motivated by permanently growing requirements for accuracy of timescales in different areas of applications of wire and wireless digital systems. Main limiters of accuracy here are the nonstationary random behavior of the clock TIE causing the model uncertainty, GPS time uncertainty due to different satellites in a view, and non Gaussian sawtooth noise induced by the commercially available GPS timing receivers owing to the principle of the 1PPS signal formation. Under such circumstances, the standard Kalman algorithm often becomes unstable and noisy, even if the sawtooth correction is used. We show that a better way is to use optimal FIR estimators, which are inherently bounded input/bounded output (BIBO) stable and more robust against temporary uncertainties and round-off errors. Among these estimators, simple unbiased FIR polynomial solutions have strong engineering features for slowly changing with time clock models. Moreover, they become optimal by large averaging horizons typically used in timekeeping. Theoretical studies of optimal FIR filtering are provided in detail. Applications are given for GPS-based measurements of crystal clocks.

1. Introduction

Although Time is a perfect eraser of any information (we have only a few evidences of ancient civilizations), modern digital information technologies rely on an extremely high accuracy of local timescales. The function of time dissemination is ordered to the Global Navigation Satellite Systems (GNSS) such as the Global Positioning System (GPS)(USA), Global Navigation System (GLONASS) (Russia), Galileo (Europe), COMPASS (China), and IRNSS (India) measuring up to world standards for signals and systems [1, 2]. Even so, precision of the disseminated time is limited with noise, which standard deviation using commercially available timing receivers is about 30 ns, can reach 10-20 ns [3] and may be improved by removal of systematic errors to no less than 3-5 ns [3, 4]. Precise correction of the local clock errors is hence not always available and optimal estimators are used. The problem we meet here is coupled with the nonstationary random behavior of the clock TIE and the GPS time temporary uncertainty caused by different satellites in a view [5]. It arises in connection with the non Gaussian sawtooth noise induced by the receiver owing to the principle of the one pulse per second (1PPS) signal formation and with the not white (colored) Gaussian noise of the clock oscillator. Under such conditions, the standard Kalman algorithm [6] cannot be used properly [7] and may become noisy and unstable [8, 9].

In this Chapter, we address an alternative approach known as the finite impulse response (FIR) filtering or weighted averaging. Contrary to the infinite impulse response (IIR) recursive structures such as the Kalman filter, the transversal FIR structures are inherently bounded input/bounded output (BIBO) stable and have better robustness against temporary uncertainties and round-off errors [10]. In the sequel, we shall show that such splendid properties make FIR estimators very attractive and even natural for applications in timescales.

2. Timescale Errors

Any clock is characterized with the time interval error (TIE) that is the instantaneous difference of the clock time and the absolute time. Because the absolute time is not available, the GPS timescale has been formed using atomic clocks. To place the GPS time close to the absolute time, measurements are permanently provided in the United States Naval Observatory (USNO) for the coordinated universal time (UTC) of the USNO master clock (MC). In order to provide an estimate of UTC (USNO MC) time derivable from a GPS signal, a set of UTC corrections is provided as part of the message signal. This broadcast message includes the time difference in whole seconds between GPS time and UTC (USNO MC). During 1996, this difference was 11 s [5]. The message also includes the rate and time difference estimate between GPS time and UTC(USNO MC) allowing for a receiver to calculate an accurate estimate of UTC(USNO MC) with the mission goal of 28 ns (1 sigma) [5]. Outside the selective availability (SA) induced for security purposes, the estimate may have an accuracy of about 10 ns in the root mean squares (RMS) sense. Practically, the USNO has been successful in predicting UTC to within about 10 ns [5]. A real-time potential uncertainty of GPS time for UTC (USNO MC) stays therefore at a level of about 14-ns [5]. The standard deviation of GPS time available precision mostly depends on random errors in different onboard clocks and different satellites in a view. It can be achieved at a level of 3-5 ns [3, 4].

2.1. GPS-based Clock Estimation and Steering

For ground applications, GPS time is disseminated via the timing receivers. In the commercially available receivers such as of the Motorola family, the 1PPS output is generated by the local time clocks (LTCs) referred to the GPS timing signals. A typical set of clock synchronization via the 1PPS signals is shown in Fig. 1.

To obtain high accuracy, all time delays featured to channels and signal propagation are compensated at the early stage. The main source of random errors that cannot be removed

GPS-based Optimal FIR Filtering and Steering of Clock Errors

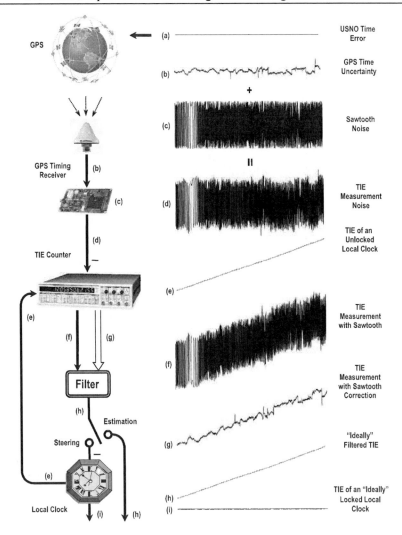

Figure 1. A typical set of GPS-based clock estimation and steering.

from the received signal below 3–5 ns [3] is associated with the GPS time precision (b) limited by different satellites in a view at their nonstationary orbits. The receiver adds the sawtooth noise (c) bounded with $\pm\Delta$, where $\Delta[\text{ns}] = \frac{10^3}{2 f_{\text{LTC}}[\text{MHz}]}$ and f_{LTC} is the frequency of the LTC oscillator. In a Motorola family of the receivers such as the SynPaQ III GPS Sensor, this frequency is chosen to be 10 MHz and the bound is thus $\Delta = 50$ ns. It would be 14 ns if $f_{\text{LTC}} = 40$ MHz.

Contrary to the error (d) in the GPS timing 1 PPS signal that is conventionally zero-mean with the uncertainty structure (b) and sawtooth noise (c), the TIE (e) of a local clock is inherently nonstationary and random, although having low intensity short-time noise. If a high resolution time interval counter is used such as the Stanford Frequency Counter SR620, then the difference between (e) and (d) represents the local clock measured TIE. The latter has the form (f) or (g) if measurements are provided with and without sawtooth (using the negative sawtooth code supplied by the receiver), respectively. To smooth excursions and

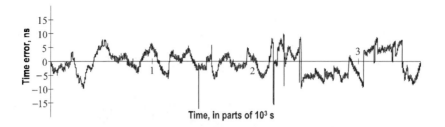

Figure 2. Typical GPS time error u_n at the receiver.

eliminate sawtooth in an optimum way, a digital estimator (Filter) of the clock TIE is used. If the filter is optimal or close to optimal in the sense of the minimum mean square error (MSE), then its output (h) ranges most closely to the actual behavior (e) and the TIE of a locked clock becomes near stationary (i) with a small amount of random departures.

2.2. Main Sources of Time Errors

Although Fig. 1 exhibits a near linear behavior of the unlocked clock TIE (e), this function is random and can be strongly nonlinear that causes difficulties in accurate estimation of clock state. The limiters of a high precision estimate are in the sawtooth (c) and in the GPS time (b). In what follows, we therefore shall recognize 1) the GPS time error caused by uncertainty and noise accumulated in the 1PPS signal from the GPS channels and 2) the sawtooth noise induced by the receiver.

2.2.1. GPS Time Error

Figure 2 shows a typical GPS time error u_n at the receiver as a difference between the TIE measurement (crystal clock) without sawtooth (g) and the actual TIE measured for the reference (cesium) standard of frequency. Systematic errors are removed. The standard deviation calculates here 3.98 ns fitting limiting facilities of the GPS time [3].

Probability density. On a long time base, the GPS time error does not exhibit periodicity (Fig. 3). As can be seen, the process histogram, apart from zero, is well approximated by the normal law. However, its peak value is typically biased (Fig. 3c and Fig. 3d) owing to various factors, including temperature effects. The process u_n is thus nonstationary and only approximately can be treated as zero-mean Gaussian and stationary in the narrow sense. In applications, such a behavior of u_n causes biased estimates of the TIE on a short time base.

Autocovariance function. The autocovariance function of the discrete-time process u_k can be calculated by

$$P_k = \frac{1}{N_m - k} \sum_{i=1}^{N_m - k} u_i u_{k+i}, \quad k = 0, 1, \ldots, N_u \ll N_m. \tag{1}$$

where N_m is a sufficiently large number of measurements. The normalized autocovariance function $\rho_{Pk} = P_k/P_0$ is shown in Fig. 4 for short correlation times (a) and large correlations times (b). For such a function, the power spectral density of u_k ranges in a narrow frequency range occupied by the local clock TIE. Therefore accurate filtering of u_k is problematic for crystal clocks. However, accuracy is higher for atomic clocks, which TIE occupies much narrower frequency range.

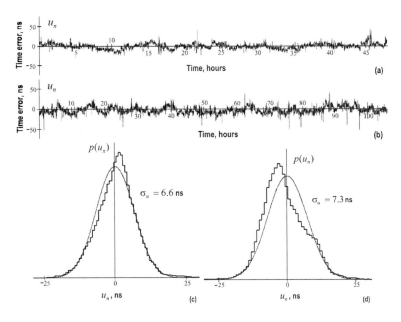

Figure 3. Typical several days measurement of the GPS time error u_n at a receiver: (a) 2 days (b) 4.5 days, (c) histogram of the process (a), and (d) histogram of the process (b).

2.2.2. Sawtooth Noise

A nuisance effect produced by GPS timing receivers is known as sawtooth. The relevant noise v_n appears in the GPS timing receiver owing to the principle of the 1 PPS signal formation having a typical structure is shown in Fig. 5. The sawtooth is caused by the nonstationary random phase of the LTC oscillator and its structure has important statistical properties discussed below.

Probability density. In low precision LTC oscillators, the white Gaussian frequency noise $\tilde{f}(t)$ dominates at the Fourier frequency f = 1 Hz associated with the 1 PPS signal. This noise has zero-mean, $E\{\tilde{f}(t)\} = 0$, and the variance $\sigma_f^2 = S_f \delta(\tau)$, where S_f is the uniform power spectral density and $\delta(t)$ is the Dirac delta function [1]. The random phase $\tilde{\varphi}(t)$ of an LTC oscillator can thus be modeled as the Wiener process [11], by solving an equation

$$\frac{d\tilde{\varphi}(t)}{dt} = 2\pi \tilde{f}(t), \quad \tilde{\varphi}(0) = \varphi_0, \qquad (2)$$

which general solution is

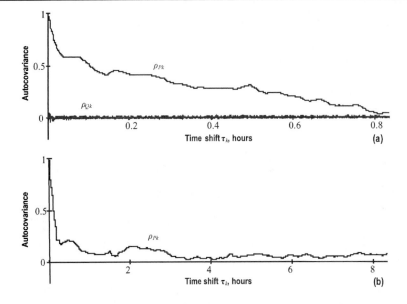

Figure 4. Typical normalized autocovariance function ρ_{Pk} of the GPS time error u_n: (a) short correlation times τ_k and (b) large correlation times τ_k; ρ_{Qk} represents the sawtooth noise.

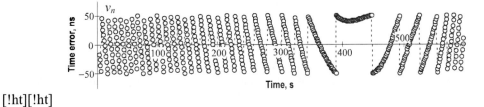

[!ht][!ht]

Figure 5. Typical sawtooth induced by the GPS timing receiver.

$$\tilde{\varphi}(t) = \varphi_0 + 2\pi \int_0^t \tilde{f}(\tau)d\tau \, . \qquad (3)$$

By multiplying (3) with $1/2\pi f_{\mathrm{LTC}}$, we go to the equation for the continuous-time noise,

$$\tilde{v}(t) = v_0 + \int_0^t \frac{\tilde{f}(\tau)}{f_{\mathrm{LTC}}} d\tau = v_0 + \int_0^t \tilde{y}(\tau)d\tau \, , \qquad (4)$$

where $\tilde{v}(t) = \tilde{\varphi}/2\pi f_{\mathrm{LTC}}$ and $v_0 = \tilde{v}(0)$. Here $\tilde{y}(t) = \tilde{f}(t)/f_{\mathrm{LTC}}$ is an instantaneous noisy fractional frequency offset with zero-mean, $E\{\tilde{y}(t)\} = 0$, and autocovariance $R_y(\tau) = h_0 f^0 \delta(\tau)$, where h_0 is a constant postulated by the IEEE Standard [12]. For the mean value v_0, the variance of $\tilde{v}(t)$ is given by

$$\sigma_{\tilde{v}}^2 = E\{[\tilde{v}(t) - v_0]^2\}$$

$$= \int_0^t \int_0^t E\{\tilde{y}(t_1)\tilde{y}(t_2)\}\mathrm{d}t_1\mathrm{d}t_2\,,$$

$$= h_0 f^0 \int_0^t \int_0^t \delta(t_2 - t_1)\mathrm{d}t_1\mathrm{d}t_2 = h_0 f^0 t\,. \tag{5}$$

Because the process $\tilde{v}(t)$ is Gaussian, its conditional nonstationary probability density function (pdf) can be written as

$$p(\tilde{v}; t|v_0) = \frac{1}{\sqrt{2\pi h_0 f^0 t}} \exp\left[-\frac{(\tilde{v} - v_0)^2}{2h_0 f^0 t}\right]\,, \tag{6}$$

where $\tilde{v}(t)$ ranges within the infinite bounds, $-\infty < \tilde{v} - v_0 < \infty$.

The principle of the 1PPS signal formation utilized in a Motorola family of the GPS timing receivers presumes changing the code of the counter if $\tilde{v}(t)$ occurs to lie beyond the bounds $\pm\Delta = \pm 1/2 f_{\mathrm{LTC}}$. This makes the modulo 2Δ time error noise $v(t)$ sawtooth (Fig. 4) and its pdf can be written as

$$p(v; t|v_0) = \frac{1}{\sqrt{2\pi h_0 f^0 t}} \sum_{k=-\infty}^{\infty} \exp\left[-\frac{(\tilde{v} - v_0 - 2\Delta k)^2}{2h_0 f^0 t}\right]\,, \tag{7}$$

where $-\Delta \leqslant v \leqslant \Delta$. An actual pdf is formed by overlapping all of the possible values of $v(t)$ available over time. For infinite t, the sum in (7) can be substituted with the integral having infinite bounds that yields

$$p(v|v_0) = \lim_{t\to\infty} p(v; t|v_0)$$

$$= \frac{1}{\sqrt{2\pi h_0 f^0 t}} \int_{-\infty}^{\infty} \exp\left[-\frac{(\tilde{v} - v_0 - 2\Delta\tau)^2}{2h_0 f^0 t}\right]\mathrm{d}\tau$$

$$= \frac{1}{2\Delta} \int_{-\infty}^{\infty} \frac{1}{\sqrt{2\pi h_0 f^0 t/4\Delta^2}} \exp\left\{-\frac{[\tau - (v_0 - \tilde{v})/2\Delta]^2}{2h_0 f^0 t/4\Delta^2}\right\}\mathrm{d}\tau\,. \tag{8}$$

Because the integrand in (8) represents the normal pdf, the integral produces unity and we finally arrive at

$$p(v) = \begin{cases} \frac{1}{2\Delta}\,, & -\Delta \leqslant v \leqslant \Delta \\ 0\,, & \text{otherwise} \end{cases}\,, \tag{9}$$

meaning that the sawtooth noise is uniformly and unconditionally distributed within the bounds $\pm\Delta$. The m-order moment and the variance associated with (9) are, respectively,

$$E\{v^m\} = \int_{-\Delta}^{\Delta} v^m p(v) dv = \frac{\Delta^m}{2(m+1)}[1+(-1)^m], \qquad (10)$$

$$\sigma_v^2 = E\{v^2\} = \frac{\Delta^2}{3}. \qquad (11)$$

A typical histogram of the sawtooth existing within the bounds ± 50 ns is shown in Fig. 5. As an be seen, a spike appears at zero owing to round-off. Certainly, the spike can be neglected when Δ is much larger than a resolution of the sawtooth correction code that is 1 ns, in modern receivers. Otherwise, the sawtooth cannot be said to be distributed uniformly.

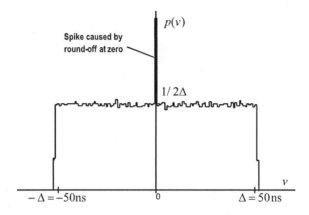

Figure 6. Typical histogram of the sawtooth within the bounds ± 50 ns.

Autocovariance function. Similarly to (1), the autocovariance function of v_n can numerically be calculated as

$$Q_k = \frac{1}{N_m - k} \sum_{i=1}^{N_m - k} v_i v_{k+i}, \quad k = 0, 1, \ldots, N_v \ll N_m, \qquad (12)$$

and we see in Fig. 7 that it is delta-shaped,

$$\rho_{Qk} = \frac{Q_k}{Q_0} \cong \begin{cases} 1, & k = 0 \\ 0, & k \neq 0 \end{cases}. \qquad (13)$$

The latter means uniformity of the power spectral density of v_n and the sawtooth can thus efficiently be filtered out. For a comparison with the GPS time error, the normalized autocovariance function ρ_{Qk} is also shown in Fig. 4a.

2.2.3. Clock Noise

The TIE of a local clock is due to the clock state dynamics and additive random constituents. Fig. 8 shows a typical evolution of the discrete time TIE of a local crystal clock, in which

Figure 7. Typical normalized autocovariance function Q_k of the sawtooth v_n.

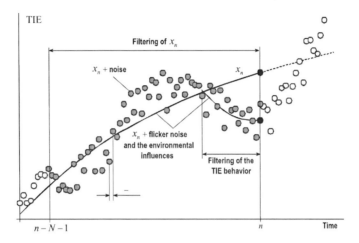

Figure 8. Typical behavior of the TIE of a local crystal clock.

one recognizes the "deterministic" trend x_n caused by the frequency offset and frequency drift. Departures from this trend are due to different constituents of the Gaussian noise acting in the clock [12], namely white phase modulation (WHPM), flicker phase modulation (FPM), white frequency modulation (WHFM), flicker frequency modulation (FFM), random walk frequency modulation (RWFM), and white random run FM noise (RRFM).

The most mystical flicker noise is hard to model. Therefore, the covariance matrix of the clock TIE noise w_{xn}, fractional frequency offset noise w_{yn}, and linear frequency drift rate noise w_{zn} is often described in the white Gaussian approximation proposed by Chaffee [13] as

$$\Psi = \tau \begin{bmatrix} q_1 + \frac{q_2 \tau^2}{3} + \frac{q_3 \tau^4}{20} & \frac{q_2 \tau}{2} + \frac{q_3 \tau^3}{8} & \frac{q_3 \tau^2}{6} \\ \frac{q_2 \tau}{2} + \frac{q_3 \tau^3}{8} & q_2 + \frac{q_3 \tau^2}{3} & \frac{q_3 \tau}{2} \\ \frac{q_3 \tau^2}{6} & \frac{q_3 \tau}{2} & q_3 \end{bmatrix}, \qquad (14)$$

where the diffusion coefficients q_1, q_2, and q_3 specify the WHFM, WRFM, and RRFM, respectively, in the averaging time τ-domain power law.

We notice that (14) is basic for the use of the Kalman filter in timescales [14, 15] and the latter is thus accurate in the white Gaussian approximation sense. There is no such a restriction on sources of noise if FIR filtering is applied [16, 17].

3. FIR Filtering of Clock States

In some applications, it is required to ascertain the clock state at a current discrete-time point n without clock steering. Measurements are thus provided in real time when the loop in Fig. 1 is open. The filter output represents the estimate of the TIE for the GPS reference timing signals and the clock state can be estimated in state space as in the following.

3.1. State Space Model of the Clock

Commonly, the clock is characterized with three states [14], namely with the time error x_n, fractional frequency offset y_n, and linear frequency drift rate z_n. For the measurement s_n of the first state x_n (Fig. 1), the real-time state space model of a clock is therefore given by

$$\mathbf{x}_n = \mathbf{A}\mathbf{x}_{n-1} + \mathbf{w}_n \,, \tag{15}$$

$$s_n = \mathbf{C}\mathbf{x}_n + e_n \,, \tag{16}$$

where $\mathbf{x}_n = [\, x_n \, y_n \, z_n \,]^T$ is the clock state vector, $\mathbf{w}_n = [\, w_{xn} \, w_{yn} \, w_{zn} \,]^T$ is the clock noise vector, and $e_n = v_n + u_n$ is the measurement noise. The matrix

$$\mathbf{A} = \begin{bmatrix} 1 & \tau & \tau^2/2 \\ 0 & 1 & \tau \\ 0 & 0 & 1 \end{bmatrix} \tag{17}$$

projects the nearest past state to the present state at n and the measurement matrix is $\mathbf{C} = [\, 1 \, 0 \, 0 \,]$. The clock noise covariance matrix

$$\mathbf{\Psi} = E\{\mathbf{w}_i \mathbf{w}_j^T\} \tag{18}$$

is supposed to have arbitrary components, although for the white Gaussian approximation it is given by (14). The variances $\sigma_u^2 = P_0$ of the GPS time errors u_n is provided with (1) and the variance σ_v^2 of the sawtooth noise v_n is given by (11).

3.1.1. Representation on a Finite Horizon

It is known that FIR filtering is provided on a horizon of $N \geqslant 2$ nearest past history points. The model (15) and (16) can fit the cases of $N = 2$ shown in Fig. 9a. However, for the most general case of $N > 2$ shown in Fig. 9b, it is in order to find forms for (15) and (16) such that all of the points from $n - N + 1$ to n are involved. That can be done using the recursively computed forward-in-time solutions. With respect to the state model (15), we thus can write

$$
\begin{aligned}
\mathbf{x}_n &= \mathbf{A}\mathbf{x}_{n-1} + \mathbf{w}_n \\
\mathbf{x}_{n-1} &= \mathbf{A}\mathbf{x}_{n-2} + \mathbf{w}_{n-1} \\
&\;\;\vdots \\
\mathbf{x}_{n-N+2} &= \mathbf{A}\mathbf{x}_{n-N+1} + \mathbf{w}_{n-N+2} \\
\mathbf{x}_{n-N+1} &= \mathbf{x}_{n-N+1} \,.
\end{aligned} \tag{19}
$$

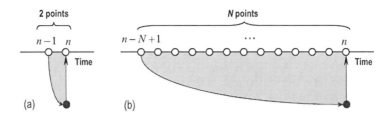

Figure 9. Averaging horizons: (a) two points and (b) N points.

Consequent substitutions of the preceding equations to the succeeding ones of (19) produce

$$\begin{aligned}
\mathbf{x}_n &= \mathbf{A}(\mathbf{A}\mathbf{x}_{n-2} + \mathbf{w}_{n-1}) + \mathbf{w}_n \\
&= \mathbf{A}^2(\mathbf{A}\mathbf{x}_{n-3} + \mathbf{w}_{n-2}) + \mathbf{A}\mathbf{w}_{n-1} + \mathbf{w}_n \\
&= \mathbf{A}^{N-1}\mathbf{x}_{n-N+1} + \mathbf{A}^{N-2}\mathbf{w}_{n-N+2} + \cdots + \mathbf{A}\mathbf{w}_{n-1} + \mathbf{w}_n, \\
\mathbf{x}_{n-1} &= \mathbf{A}(\mathbf{A}\mathbf{x}_{n-3} + \mathbf{w}_{n-2}) + \mathbf{w}_{n-1} \\
&= \mathbf{A}^2(\mathbf{A}\mathbf{x}_{n-4} + \mathbf{w}_{n-3}) + \mathbf{A}\mathbf{w}_{n-2} + \mathbf{w}_{n-1} \\
&= \mathbf{A}^{N-2}\mathbf{x}_{n-N+1} + \mathbf{A}^{N-3}\mathbf{w}_{n-N+2} + \cdots + \mathbf{A}\mathbf{w}_{n-2} + \mathbf{w}_{n-1}, \\
&\vdots \\
\mathbf{x}_{n-N+2} &= \mathbf{A}\mathbf{x}_{n-N+1} + \mathbf{w}_{n-N+2}, \\
\mathbf{x}_{n-N+1} &= \mathbf{x}_{n-N+1}
\end{aligned} \qquad (20)$$

that can be rewritten in a short matrix form as

$$\mathbf{X}_N = \mathbf{A}_N \mathbf{x}_{n-N+1} + \mathbf{B}_N \mathbf{N}_N, \qquad (21)$$

where the $3N \times 1$ clock state vector is given by

$$\mathbf{X}_N = \begin{bmatrix} \mathbf{x}_n^T & \mathbf{x}_{n-1}^T & \cdots & \mathbf{x}_{n-N+1}^T \end{bmatrix}^T, \qquad (22)$$

and the $3N \times 3$ transition matrix is

$$\mathbf{A}_r = \begin{bmatrix} (\mathbf{A}^{r-1})^T & (\mathbf{A}^{r-2})^T & \cdots & (\mathbf{A}^{r-N})^T \end{bmatrix}^T, \qquad (23)$$

in which

$$\mathbf{A}^i = \begin{bmatrix} 1 & \tau i & \tau^2 i^2/2 \\ 0 & 1 & \tau \\ 0 & 0 & 1 \end{bmatrix}, \qquad (24)$$

$\mathbf{A}^0 = \mathbf{I}$ is identity, and $\mathbf{A}^i$ has all components equal to zero if $i < 0$.

The $3N \times 1$ clock noise vector is described with

$$\mathbf{N}_N = \begin{bmatrix} \mathbf{w}_n^T & \mathbf{w}_{n-1}^T & \cdots & \mathbf{w}_{n-N+1}^T \end{bmatrix}^T \qquad (25)$$

and the $3N \times 3N$ input matrix has two equal forms of

$$
\mathbf{B}_N = \begin{bmatrix} \mathbf{I} & \mathbf{A} & \dots & \mathbf{A}^{N-2} & \mathbf{A}^{N-1} \\ \mathbf{0} & \mathbf{I} & \dots & \mathbf{A}^{N-3} & \mathbf{A}^{N-2} \\ \vdots & \vdots & \ddots & \vdots & \vdots \\ \mathbf{0} & \mathbf{0} & \dots & \mathbf{I} & \mathbf{A} \\ \mathbf{0} & \mathbf{0} & \dots & \mathbf{0} & \mathbf{I} \end{bmatrix} \tag{26}
$$

$$
= \begin{bmatrix} \mathbf{A}_1 & \mathbf{A}_2 & \dots & \mathbf{A}_{N-1} & \mathbf{A}_N \end{bmatrix}, \tag{27}
$$

where $\mathbf{I}$ is an identity matrix of proper dimensions and $\mathbf{0}$ is a relevant matrix with all components equal to zero. Note that the initial state vector $\mathbf{x}_{n-N+1}$ is given in (19) and (20) exactly, although it is randomly valued. Therefore, the noise vector $\mathbf{w}_{n-N+1}$ in (25) is always zero-valued.

In a like manner, the measurement equation (16) can now be represented on a horizon of N past points as

$$
\begin{aligned}
s_n &= \mathbf{C}\mathbf{x}_n + e_n \\
s_{n-1} &= \mathbf{C}\mathbf{x}_{n-1} + e_{n-1} \\
&\;\;\vdots \\
s_{n-N+2} &= \mathbf{C}\mathbf{x}_{n-N+2} + e_{n-N+2} \\
s_{n-N+1} &= \mathbf{C}\mathbf{x}_{n-N+1} + e_{n-N+1}.
\end{aligned} \tag{28}
$$

Substituting the clock states taken from (20) allows us to write

$$
\begin{aligned}
s_n &= \mathbf{C}\mathbf{A}^{N-1}\mathbf{x}_{n-N+1} + \mathbf{C}\mathbf{w}_n + \mathbf{C}\mathbf{A}\mathbf{w}_{n-1} + \mathbf{C}\mathbf{A}^2\mathbf{w}_{n-2} \\
&\quad + \dots + \mathbf{C}\mathbf{A}^{N-2}\mathbf{w}_{n-N+2} + e_n \\
s_{n-1} &= \mathbf{C}\mathbf{A}^{N-2}\mathbf{x}_{n-N+1} + \mathbf{C}\mathbf{w}_{n-1} + \mathbf{C}\mathbf{A}\mathbf{w}_{n-2} + \mathbf{C}\mathbf{A}^2\mathbf{w}_{n-3} \\
&\quad + \dots + \mathbf{C}\mathbf{A}^{N-3}\mathbf{w}_{n-N+2} + e_{n-1} \\
&\;\;\vdots \\
s_{n-N+2} &= \mathbf{C}\mathbf{A}\mathbf{x}_{n-N+1} + \mathbf{C}\mathbf{w}_{n-N+2} + e_{n-N+2} \\
s_{n-N+1} &= \mathbf{C}\mathbf{x}_{n-N+1} + e_{n-N+1}
\end{aligned} \tag{29}
$$

that can also be rewritten in a short matrix form as

$$
\mathbf{S}_N = \mathbf{C}_N\mathbf{x}_{n-N+1} + \mathbf{G}_N\mathbf{N}_N + \mathbf{V}_N, \tag{30}
$$

where the $N \times 1$ measurement vector of the TIE is

$$
\mathbf{S}_N = \begin{bmatrix} s_n\, s_{n-1} \dots s_{n-N+1} \end{bmatrix}^T \tag{31}
$$

and the $N \times 3$ measurement matrix has the forms of, if to set $k = N$,

$$\mathbf{C}_r = \begin{bmatrix} \mathbf{CA}^{r-1} \\ \mathbf{CA}^{r-2} \\ \vdots \\ \mathbf{CA}^{r-N} \end{bmatrix} = \begin{bmatrix} (\mathbf{A}^{r-1})_1 \\ (\mathbf{A}^{r-2})_1 \\ \vdots \\ (\mathbf{A}^{r-N})_1 \end{bmatrix},$$

$$\mathbf{C}_N = \begin{bmatrix} 1 & \tau(N-1) & \tau^2(N-1)^2/2 \\ 1 & \tau(N-2) & \tau^2(N-2)^2/2 \\ \vdots & \vdots & \vdots \\ 1 & \tau & \tau^2/2 \\ 1 & 0 & 0 \end{bmatrix}. \tag{32}$$

Here and in the following, $(\mathbf{Z})_i$ means the i-row of a matrix $\mathbf{Z}$.

The $N \times 1$ measurement noise vector is represented with

$$\mathbf{V}_N = \begin{bmatrix} e_n \, e_{n-1} \, \cdots \, e_{n-N+1} \end{bmatrix}^T, \tag{33}$$

and an auxiliary $N \times 3N$ matrix takes one of the forms of

$$\mathbf{G}_N = \begin{bmatrix} \mathbf{C} & \mathbf{CA} & \cdots & \mathbf{CA}^{N-2} & \mathbf{CA}^{N-1} \\ 0 & \mathbf{C} & \cdots & \mathbf{CA}^{N-3} & \mathbf{CA}^{N-2} \\ \vdots & \vdots & \ddots & \vdots & \vdots \\ 0 & 0 & \cdots & \mathbf{C} & \mathbf{CA} \\ 0 & 0 & \cdots & 0 & \mathbf{C} \end{bmatrix} \tag{34}$$

$$= \begin{bmatrix} (\mathbf{I})_1 & (\mathbf{A})_1 & \cdots & (\mathbf{A}^{N-2})_1 & (\mathbf{A}^{N-1})_1 \\ 0 & (\mathbf{I})_1 & \cdots & (\mathbf{A}^{N-3})_1 & (\mathbf{A}^{N-2})_1 \\ \vdots & \vdots & \ddots & \vdots & \vdots \\ 0 & 0 & \cdots & (\mathbf{I})_1 & (\mathbf{A})_1 \\ 0 & 0 & \cdots & 0 & (\mathbf{I})_1 \end{bmatrix}$$

$$= \begin{bmatrix} \mathbf{C}_1 & \mathbf{C}_2 & \cdots & \mathbf{C}_{N-1} & \mathbf{C}_N \end{bmatrix}. \tag{35}$$

So, the state space model (15) and (16) is now represented by the following equations matched with the averaging horizon of N nearest past points,

$$\mathbf{X}_N = \mathbf{A}_N \mathbf{x}_{n-N+1} + \mathbf{B}_N \mathbf{N}_N, \tag{36}$$

$$\mathbf{S}_N = \mathbf{C}_N \mathbf{x}_{n-N+1} + \mathbf{G}_N \mathbf{N}_N + \mathbf{V}_N, \tag{37}$$

where all of the matrices. $\mathbf{A}_N$, $\mathbf{B}_N$, $\mathbf{C}_N$, and $\mathbf{G}_N$, are composed by the components of the matrix $\mathbf{A}$. The model involves the clock history starting at the initial state $\mathbf{x}_{n-N+1}$ and ending with the current state $\mathbf{x}_n$. To provide FIR filtering of x_n, the discrete convolution can now be used straightforwardly.

3.2. Optimal FIR Filtering of the TIE

Utilizing N measurement points from $n - N + 1$ to n, the FIR filtering estimate $\hat{x}_n$ of the TIE can be obtained at n via (36) and (37) by the discrete convolution as follows

$$
\begin{aligned}
\hat{x}_n &= \sum_{i=0}^{N-1} h_{li} s_{n-i} && (38) \\
&= \mathbf{W}_l^T \mathbf{S}_N && (39) \\
&= \mathbf{W}_l^T (\mathbf{C}_N \mathbf{x}_{n-N+1} + \mathbf{G}_N \mathbf{N}_N + \mathbf{V}_N), && (40)
\end{aligned}
$$

where h_{li} is the l-degree FIR filter gain [17]. We shall further show that the filter degree plays an important role in optimal FIR filtering.

The FIR filter gain matrix is given on a horizon of N points as

$$
\mathbf{W}_l = \begin{bmatrix} h_{l0} & h_{l1} & \dots & h_{l(N-1)} \end{bmatrix}^T. \tag{41}
$$

To optimize $\mathbf{W}_l$ in the sense of the minimum MSE, the cost function can be written using (40) as

$$
\begin{aligned}
J &= E(x_n - \hat{x}_n)^2 \\
&= E[x_n - \mathbf{W}_l^T (\mathbf{C}_N \mathbf{x}_{n-N+1} + \mathbf{G}_N \mathbf{N}_N + \mathbf{V}_N)]^2, \tag{42}
\end{aligned}
$$

where E means an average of the succeeding expression.

To minimize (42), one can equate to zero the gradient of J with respect to the matrix $\mathbf{W}_l$, let $\mathbf{W}_l$ to be optimal, $\mathbf{W}_{l0}$, and arrive at the orthogonality condition [18],

$$
\begin{aligned}
& E[x_n - \mathbf{W}_{l0}^T (\mathbf{C}_N \mathbf{x}_{n-N+1} + \mathbf{G}_N \mathbf{N}_N + \mathbf{V}_N)] \\
& (\mathbf{C}_N \mathbf{x}_{n-N+1} + \mathbf{G}_N \mathbf{N}_N + \mathbf{V}_N)^T = 0. \tag{43}
\end{aligned}
$$

Further substituting x_n with the model fitting TIE behavior (Fig. 8) needs a special discussion. Below, we recognize several special cases suitable for clocks.

- If all of the noise components in the model and in the measurement have to be filtered out in the estimation of the clock states, then x_n must be substituted with its deterministic part taken from the first row of (36),

$$
x_n = (\mathbf{A}^{N-1})_1 \mathbf{x}_{n-N+1}, \tag{44}
$$

where the first rows of $\mathbf{A}^{N-1}$ is

$$
(\mathbf{A}^{N-1})_1 = \begin{bmatrix} 1 & \tau(N-1) & \tau^2(N-1)^2/2 \end{bmatrix}.
$$

This case requires large averaging horizons as shown in Fig. 8.

- In GPS-based synchronization, the slow process caused by flicker noise and the environmental factors, such as temperature, is typically tracked along with the first clock state. To enable, the noise matrix $\mathbf{N}_N$ may be separated into the matrix $\bar{\mathbf{N}}_N$ containing the slowly changing with time components and $\tilde{\mathbf{N}}_N$ containing the white noise components. Then $\mathbf{N}_N$ can be substituted in (43) with $\tilde{\mathbf{N}}_N$ and (44) used. Note that an averaging horizon in this case would be shorter, as shown in Fig. 8. Alternatively, one can substitute x_n with

$$x_n = (\mathbf{A}^{N-1})_1 \mathbf{x}_{n-N+1} + (\mathbf{B}_N)_1 \bar{\mathbf{N}}_N \,,$$

and find an optimal gain for this model.

- In order to estimate the Allan variance via noisy measurements, the clock noise structure has to be saved completely in the filter output and the measurement noise filtered out. In this special case, the complete first state noise model can be used as in [16],

$$x_n = (\mathbf{A}^{N-1})_1 \mathbf{x}_{n-N+1} + (\mathbf{B}_N)_1 \mathbf{N}_N \,.$$

In what follows, we will mostly provide state estimation and clock synchronization using the model (44).

Substituting (44) to (41) and saving the complete clock noise structure in $\mathbf{N}_N$ give

$$E[(\mathbf{A}^{N-1})_1 \mathbf{x}_{n-N+1} - \mathbf{W}_{l0}^T (\mathbf{C}_N \mathbf{x}_{n-N+1} + \mathbf{G}_N \mathbf{N}_N + \mathbf{V}_N)]$$
$$(\mathbf{C}_N \mathbf{x}_{n-N+1} + \mathbf{G}_N \mathbf{N}_N + \mathbf{V}_N)^T = 0 \,. \tag{45}$$

Next transformations can be provided if we recall that the random processes in the set (Fig. 1) are mutually independent and uncorrelated; that is,

$$E\{\mathbf{x}_{n-N+1}\mathbf{N}_N^T\} = E\{\mathbf{N}_N \mathbf{x}_{n-N+1}^T\} = \mathbf{0} \,,$$
$$E\{\mathbf{x}_{n-N+1}\mathbf{V}_N^T\} = E\{\mathbf{V}_N \mathbf{x}_{n-N+1}^T\} = \mathbf{0} \,,$$
$$E\{\mathbf{V}_N \mathbf{N}_N^T\} = E\{\mathbf{N}_N \mathbf{V}_N^T\} = \mathbf{0} \,.$$

If also to introduce the noise covariance matrices

$$\mathbf{\Psi}_N = E\{\mathbf{N}_N \mathbf{N}_N^T\} \,,$$
$$\mathbf{\Phi}_V = E\{\mathbf{V}_N \mathbf{V}_N^T\}$$

and the mean square initial state at $n - N + 1$,

$$\mathbf{R}_0 = E\{\mathbf{x}_{n-N+1}\mathbf{x}_{n-N+1}^T\} \,,$$

then averaging of (45) will lead to the exact optimal FIR filter gain transpose matrix [19]

$$\mathbf{W}_{l0}^T = (\mathbf{A}^{N-1})_1 \mathbf{R}_0 \mathbf{C}_N^T (\mathbf{Z}_0 + \mathbf{Z}_N + \mathbf{\Phi}_V)^{-1} \,, \tag{46}$$

where $\mathbf{Z}_0 = \mathbf{C}_N \mathbf{R}_0 \mathbf{C}_N^T$ and $\mathbf{Z}_N = \mathbf{G}_N \mathbf{\Psi}_N \mathbf{G}_N^T$.

Another form of $\mathbf{W}_{l0}^T$ was shown in [16] for the estimation of the Allan variance via noisy measurements, when only the measurement noise is filtered out. Following the above-given analysis and [16], the gain matrix can easily be modified for clock steering, when only the white noise components are filtered out.

One can observe that (46) requires the initial state error matrix $\mathbf{R}_0$ as well as the clock noise covariance matrix $\mathbf{\Psi}_V$ and the measurement noise covariance matrix $\mathbf{\Phi}_V$. Below, we shall show that, by large $N \gg 1$ typical for clock problems, the gain (44) is simplified substantially, but still remains optimal.

3.2.1. Large Averaging Horizon $N \gg 1$

The main problem in (46) arises in the computation of the inverse matrix $\mathbf{Z}^{-1} = (\mathbf{Z}_0 + \mathbf{Z}_N + \mathbf{\Phi}_V)^{-1}$. This matrix has $N \times N$ dimensions and must exist. On the other hand, it is known that the clock states are inherently slowly changing with time owing to high quality factors featured to clock oscillators, low sensitivity to the environment, and small aging. Therefore, estimation of the clock states is provided in different sense on different time intervals depending on applications:

- A horizon of $10^2 \dots 10^3$ points typically is enough for the synchronization of crystal and rubidium clocks.

- About 10^5 s are required for filtering and prediction in atomic clocks.

- Large intervals of about 10^7 s are employed to estimate the clock noise structure via the Allan variance.

This observation assures us that an averaging horizon is always large in applications to clocks, $N \gg 1$. To figure out whether a large N resides in simplification of (46), an analysis of components in $\mathbf{Z}_0$, $\mathbf{Z}_N$, and $\mathbf{\Phi}_V$ is needed for typical clock errors. Because the 1×1 components, Z_{011} and Z_{N11}, have weight N, they dominate all others. Thus, a comparison of Z_{011}, Z_{N11}, and Φ_{V11} may give us an answer. Below, an analysis is provided for clocks with oven controlled crystal oscillators (OCXOs).

Initial Clock Error. The initial clock state vector $\mathbf{x}_{n-N+1}$ is determined randomly within allowed bounds. If we suppose that its components are mutually uncorrelated then $\mathbf{R}_0$ would be diagonal with the mean square states $\langle x_{n-N+1}^2 \rangle$, $\langle y_{N-n+1}^2 \rangle$, and $\langle z_{N-n+1}^2 \rangle$ as components. From $\mathbf{Z}_0 = \mathbf{C}_N \mathbf{R}_0 \mathbf{C}_N^T$ we thus have

$$
\begin{aligned}
Z_{011} &= \langle x_{n-N+1}^2 \rangle + \tau^2 (N-1)^2 \langle y_{N-n+1}^2 \rangle \\
&\quad + \frac{\tau^4 (N-1)^4}{4} \langle z_{N-n+1}^2 \rangle .
\end{aligned}
$$

The initial clock error with equal probability takes a value from 0 to τ. For $\tau = 1$ s, its mean square value is thus $\langle x_{n-N+1}^2 \rangle \cong 0.333$ s^2 [11]. Reasoning similarly for typical OCXO-based clocks, one may let $\langle y_{n-N+1}^2 \rangle \cong 10^{-22}$ and $\langle z_{n-N+1}^2 \rangle \cong 4 \times 10^{-30}$ s^{-2}. Then this component becomes

$$Z_{011}, s^2 \cong 0.333 + 10^{-22} N^2 + 10^{-30} N^4, \tag{47}$$

meaning that the first term dominates if $N \leqslant 10^6$. Hence, $Z_{011} \cong 0.333$ s^2.

Clock Noise. In white Gaussian approximation, the 1×1 component of $\mathbf{Z}_N = \mathbf{G}_N \mathbf{\Psi}_N \mathbf{G}_N^T$ can be written as

$$
\begin{aligned}
Z_{N11} &= \sigma_x^2 N + \sigma_y^2 \tau^2 \sum_{i=0}^{N-1} i + \sigma_z^2 \frac{\tau^4}{4} \sum_{i=0}^{N-1} i^2 \\
&\cong \sigma_x^2 N + \sigma_y^2 \tau^2 \frac{N^2}{2} + \sigma_z^2 \tau^4 \frac{N^3}{12},
\end{aligned}
$$

where σ_x^2, σ_y^2, and σ_z^2 are noise variances of the relevant clock states. We neglect the third term, because σ_z^2 is small and not specified for the OCXOs. We also allow $\sigma_x^2 \cong \tau^2 N \sigma_y^2$ in the Wiener process approximation. Then, for a typical value $\sigma_y^2 = 2 \times 10^{-12}/3$, we find

$$Z_{N11} \cong \sigma_y^2 \tau^2 \frac{3N^2}{2} \cong 10^{-12} \tau^2 N^2. \tag{48}$$

Measurement Noise. The 1×1 component of $\mathbf{\Phi}_V$ is specified by the measurement noise variance, $\Phi_{V11} = \sigma_v^2$. The sawtooth error induced by the GPS timing receiver is uniformly distributed from -50 ns to 50 ns with the variance $\sigma_v^2 \cong 833$ ns^2 [11]. For this error we have $\Phi_{V11} \cong 8.3 \times 10^{-16}$ s^2.

Some Important Generalizations. One can now evaluate the 1×1 component of $\mathbf{Z}$ by

$$
\begin{aligned}
Z_{11} &\cong \langle x_{n-N+1}^2 \rangle + \sigma_y^2 \tau^2 \frac{3N^2}{2} + \sigma_v^2 \\
&\cong (0.333 + 10^{-12} N^2 + 8.3 \times 10^{-16}) \, \text{s}^2
\end{aligned}
\tag{49}
$$

that instantly lead to an important inference. For $N < 10^6$ (averaging time does not exceed 11 days), the 1×1 component of $\mathbf{Z}_0$ dominates in Z_{11}. The same can be said about the remaining components of $\mathbf{Z}_0$. Moreover, it should hold true for atomic clocks, in which all of the errors are diminished and $\langle x_{n-N+1}^2 \rangle$ still must dominate.

Such fundamental observations allows for a dramatic simplification in (46). In fact, neglecting $\mathbf{Z}_N$ and $\mathbf{\Phi}_V$ degenerates (46) to the form of

$$\mathbf{W}_{l0}^T = (\mathbf{A}^{N-1})_1 \mathbf{R}_0 \mathbf{C}_N^T (\mathbf{C}_N \mathbf{R}_0 \mathbf{C}_N^T)^{-1}. \tag{50}$$

By multiplying the both sides of (50) with $\mathbf{C}_N \mathbf{R}_0 \mathbf{C}_N^T$,

$$\mathbf{W}_{l0}^T \mathbf{C}_N \mathbf{R}_0 \mathbf{C}_N^T = (\mathbf{A}^{N-1})_1 \mathbf{R}_0 \mathbf{C}_N^T,$$

and then discarding $\mathbf{R}_0 \mathbf{C}_N^T$, one arrives at the deadbeat (or unbiasedness) constraint

$$\mathbf{W}_{l0}^T \mathbf{C}_N = (\mathbf{A}^{N-1})_1. \tag{51}$$

188   Yuriy Shmaliy

Because (51) does not depend on the initial conditions, one may deduce that the initial state covariance $\mathbf{R}_0$ can be arbitrary in (50), although such that the inverse $(\mathbf{C}_N\mathbf{R}_0\mathbf{C}_N^T)^{-1}$ exists.

> **Resume:** *On large averaging horizons, $N \gg 1$, GPS-based optimal filtering of clock state is provided with the unbiased FIR filter produced by the deadbeat constraint (49).*

The constraint (51) has been exploited in the literature to derive unbiased FIR estimators of different kinds. An unbiased optimal FIR filter was derived in [9] for receding horizon control and a purely unbiased solution was found for the same problem in [20]. For real-time signal processing at n, the unbiased FIR filter was proposed in [21] and generalized in [17]. The p-step predictive unbiased FIR filter was addressed in [22].

3.2.2. Estimate Variance

The variance of the estimate $\hat{x}_n$ can now be found by analyzing the MSE (40) and substituting x_n with (42),

$$
\begin{aligned}
J &= E(x_n - \hat{x}_n)^2 \\
&= E[x_n - \mathbf{W}_l^T(\mathbf{C}_N\mathbf{x}_{n-N+1} + \mathbf{G}_N\mathbf{N}_N + \mathbf{V}_N)]^2 \\
&= E[(\mathbf{A}^{N-1})_1 x_{n-N+1} - \mathbf{W}_l^T(\mathbf{C}_N\mathbf{x}_{n-N+1} + \mathbf{G}_N\mathbf{N}_N + \mathbf{V}_N)]^2 . \quad (52)
\end{aligned}
$$

Embedded the unbiasedness (51) and accounted for the commutativity of $\mathbf{W}_l^T\mathbf{G}_{Nk}\mathbf{N}_N = (\mathbf{G}_{Nk}\mathbf{N}_N)^T\mathbf{W}_l$ and $\mathbf{W}_l^T\mathbf{V}_N = \mathbf{V}_N^T\mathbf{W}_l$, the MSE (52) represents the variance

$$
\begin{aligned}
\sigma_x^2 &= E(\mathbf{W}_{l0}^T\mathbf{G}_N\mathbf{N}_N + \mathbf{W}_{l0}^T\mathbf{V}_N)^2 \\
&= E\{\mathbf{W}_{l0}^T\mathbf{G}_N\mathbf{N}_N\mathbf{W}_{l0}^T\mathbf{G}_N\mathbf{N}_N\} + E\{\mathbf{W}_{l0}^T\mathbf{V}_N\mathbf{W}_{l0}^T\mathbf{V}_N\} \\
&= \mathbf{W}_{l0}^T\mathbf{G}_N E\{\mathbf{N}_N\mathbf{N}_N^T\}\mathbf{G}_N^T\mathbf{W}_{l0} + \mathbf{W}_{l0}^T E\{\mathbf{V}_N\mathbf{V}_N^T\}\mathbf{W}_{l0} \\
&= \mathbf{W}_{l0}^T\left(\mathbf{G}_N E\{\mathbf{N}_N\mathbf{N}_N^T\}\mathbf{G}_N^T + E\{\mathbf{V}_N\mathbf{V}_N^T\}\right)\mathbf{W}_{l0} \\
&= \mathbf{W}_{l0}^T\left(\mathbf{G}_N\boldsymbol{\Psi}_N\mathbf{G}_N^T + \boldsymbol{\Phi}_V\right)\mathbf{W}_l \\
&= \mathbf{W}_{l0}^T\left(\mathbf{Z}_N + \boldsymbol{\Phi}_V\right)\mathbf{W}_l . \quad (53)
\end{aligned}
$$

An evaluation of σ_x^2 needs now further analysis.

We have already calculated the 1×1 component of $\mathbf{Z}_N$ with (48). Similarly, the remaining components may be ascertained for (53). However, the constituents of clock noise are correlated and, even in the Gaussian approximation (14), the matrix $\boldsymbol{\Psi}$ has no zero components. Therefore, most generally, the matrix $\mathbf{Z}_N$ has all the components not equal to zero. This special topic is avoided in this Chapter for further consideration.

Now recall that the sawtooth component v_n of e_n is delta-correlated (Fig. 7) and the GPS time error u_n not (Fig. 4). Moreover, the sawtooth noise (Fig. 1c) dominates the GPS time error (Fig. 1d) and we can approximately let $e_n \cong v_n$. The matrix $\boldsymbol{\Phi}_N$ thus would be diagonal with the measurement noise variance σ_v^2 as components.

Referring to the difficulties with $\mathbf{Z}_N$ and neglecting u_n, we can now evaluate σ_x^2 for tracking the TIE behavior shown in Fig. 8 via GPS-based measurements with sawtooth by

$$
\begin{aligned}
\sigma_x^2 &= \mathbf{W}_l^T \operatorname{diag} \underbrace{\left(\sigma_v^2 \ \sigma_v^2 \ \dots \ \sigma_v^2 \right)}_{N} \mathbf{W}_l \\
&= \sigma_v^2 \mathbf{W}_l^T \mathbf{W}_l && (54) \\
&= \sigma_v^2 g_l , && (55)
\end{aligned}
$$

where the noise power gain (NG) [23] is specified by

$$
g_l = \mathbf{W}_l^T \mathbf{W}_l . \tag{56}
$$

In the following sections, we shall use this measure (56) to evaluate the estimate variance and ascertain the filter stability.

3.3.   Unbiased FIR Filtering of the TIE

The unbiased FIR filter can now be derived from the deadbeat constraint (49) as follows. First, for clarity, represent (49) with

$$
\begin{bmatrix} h_{l0} \ h_{l1} \ \dots \ h_{l(N-1)} \end{bmatrix}
\begin{bmatrix}
1 & \tau(N-1) & \frac{\tau^2(N-1)^2}{2} \\
1 & \tau(N-2) & \frac{\tau^2(N-2)^2}{2} \\
\vdots & \vdots & \vdots \\
1 & \tau & \frac{\tau^2}{2} \\
1 & 0 & 0
\end{bmatrix}
$$

$$
= \begin{bmatrix} 1 & \tau(N-1) & \frac{\tau^2(N-1)^2}{2} \end{bmatrix} . \tag{57}
$$

Now, equate the components of the row matrices in the left-hand and right-hand sides of (57), account the first identity in the remaining ones, and arrive at the fundamental properties of the unbiased FIR filters [17]:

$$
\sum_{i=0}^{N-1} h_{li} = 1 , \tag{58}
$$

$$
\sum_{i=0}^{N-1} h_{li} i^u = 0 , \quad 1 \leqslant u \leqslant l . \tag{59}
$$

These properties can be rewritten in a matrix form as

$$
\mathbf{W}_{l0}^T \mathbf{V} = \mathbf{J}^T , \tag{60}
$$

where $\mathbf{J} = [1 \ 0 \ \dots \ 0]^T$ is of $N \times 1$ dimensions and $\mathbf{V}$ is the Vandermonde matrix

$$
\mathbf{V} = \begin{bmatrix} 1 & 0 & 0 \\ 1 & 1 & 1 \\ 1 & 2 & 2^2 \\ \vdots & \vdots & \vdots \\ 1 & N-1 & (N-1)^2 \end{bmatrix}. \tag{61}
$$

To solve (60), take into consideration that the inverse $(\mathbf{V}^T\mathbf{V})^{-1}$ always exists, multiply the right-hand side of (60) with the identity matrix $(\mathbf{V}^T\mathbf{V})^{-1}\mathbf{V}^T\mathbf{V}$, discard $\mathbf{V}$ from both sides, and finally arrive at the fundamental solution

$$
\mathbf{W}_{l0}^T = \mathbf{J}^T(\mathbf{V}^T\mathbf{V})^{-1}\mathbf{V}^T, \tag{62}
$$

representing the unbiased FIR filter gain transpose matrix. The gain (62) can be used in (39) to find the estimate that would be unbiased for any N and optimal for $N \gg 1$.

3.3.1. Polynomial FIR Filter Gain

Although (62) is a simple solution, there is an inconvenience in using the Vandermonde matrix that has large dimensions when N is large. On the other hand, we still have no idea about the gain function h_{li}. The inconvenience can be circumvented if we invoke the fundamentals of the Kalman-Bucy filter theory [6].

> **Remark:** *It is known from the Kalman-Bucy filter theory that the order of the optimal (and so unbiased) filter is the same as that of the system.*

In accordance with this theory, the k-state of the clock characterized with K states can optimally (and unbiasedly) be filtered with the $l = K - k$ degree[1] FIR filter [17]. In other words:

- The first state, $k = 1$, of the 1-state clock model, $K = 1$, (near ideal clock) is filtered with the uniform gain (simple averaging), $l = 0$,

- The first state, $k = 1$, of the two-state model, $K = 2$, (atomic clocks) is filtered with the linear (ramp) gain, $l = 1$,

- The first state, $k = 1$, of the three-state model, $K = 3$, is filtered with the quadratic gain, $l = 2$.

Most generally [17], one may represent the unbiased FIR filter gain with the l-degree polynomial

$$
h_{li} = \sum_{j=0}^{l} a_{jl} i^j, \tag{63}
$$

where the coefficients a_{jl} still need to be determined. For the three-state clock model, $l = 2$, we thus have

---

[1]Unbiasedness is also achieved with the redundant filter degree, $l > K - k$, although with larger noise.

$$h_{2i} = a_{02} + a_{12}i + a_{22}i^2 \tag{64}$$

Substituting (64) to (62) and rearranging the terms lead to the linear equation matrix

$$\mathbf{J} = \mathbf{D}\mathbf{\Upsilon}, \tag{65}$$

$$\begin{bmatrix} 1 \\ 0 \\ 0 \end{bmatrix} = \begin{bmatrix} d_0 & d_1 & d_2 \\ d_1 & d_2 & d_3 \\ d_2 & d_3 & d_4 \end{bmatrix} \begin{bmatrix} a_{02} \\ a_{12} \\ a_{22} \end{bmatrix} \tag{66}$$

where, most generally,

$$\mathbf{J} = \underbrace{\begin{bmatrix} 1 & 0 & \dots & 0 \end{bmatrix}}_{K-k+1}^T, \tag{67}$$

$$\mathbf{\Upsilon} = \underbrace{\begin{bmatrix} a_{0(K-k)} & a_{1(K-k)} & \cdots & a_{(K-k)(K-k)} \end{bmatrix}}_{K-k+1}^T, \tag{68}$$

and a short symmetric matrix $\mathbf{D}$ is specified by the Vandermonde matrix (61) as

$$\mathbf{D} = \mathbf{V}^T \mathbf{V} = \begin{bmatrix} d_0 & d_1 & \dots & d_l \\ d_1 & d_2 & \dots & d_{l+1} \\ \vdots & \vdots & \ddots & \vdots \\ d_l & d_{l+1} & \dots & d_{2l} \end{bmatrix}. \tag{69}$$

The component in (69) is defined by

$$d_m = \sum_{i=0}^{N-1} i^m, \quad m = 0, 1, \dots 2l, \tag{70}$$

$$= \frac{1}{m+1} [B_{m+1}(N) - B_{m+1}], \tag{71}$$

where $B_n(x)$ is the Bernoulli polynomial and $B_n = B_n(0)$ is the Bernoulli coefficient. An analytic solution of (65) gives

$$a_{jl} = (-1)^j \frac{M_{(j+1)1}}{|\mathbf{D}|}, \tag{72}$$

where $\mathbf{D}$ is the determinant and $M_{(j+1)1}$ is the minor of (69).

Determined a_{jl}, the unbiased FIR filter gain can be represented in the polynomial form (63) of degree $l = K - 1$, corresponding to the 1-state (TIE) of the K-state clock model.

3.3.2. Unique Polynomial FIR Filter Gains

It is now just a matter of simple mathematical manipulations to use (72) and derive the unique unbiased FIR filter polynomial gains for typical clock models. Although precise clocks are described with $1 \leqslant K \leqslant 3$, we allow $K = 4$ for low-precision clocks, and find the gains for $0 \leqslant l \leqslant 3$ [17]:

$$h_{0i} = \frac{1}{N}, \tag{73}$$

$$h_{1i} = \frac{2(2N-1)-6i}{N(N+1)}, \tag{74}$$

$$h_{2i} = \frac{3(3N^2-3N+2)-18(2N-1)i+30i^2}{N(N+1)(N+2)}, \tag{75}$$

$$h_{3i} = \frac{8(2N^3-3N^2+7N-3)-20(6N^2-6N+5)i}{N(N+1)(N+2)(N+3)}$$
$$+\frac{120(2N-1)i^2-140i^3}{N(N+1)(N+2)(N+3)}. \tag{76}$$

Figure 10 illustrates (73)–(76) for $N \gg 1$. As can be observed, the one-state clock

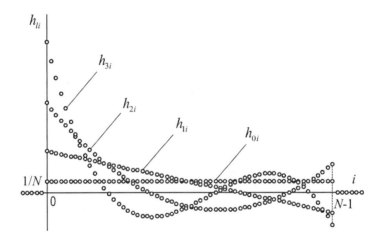

Figure 10. Unique polynomial gains of the low-degree unbiased FIR filters for the clock model.

model ($K = 1$) is filtered with the uniform gain (73) associated with simple averaging. That is expected, because simple averaging is optimal among all other filters in the sense of the minimum produced noise. It is a unique solution for the one-state model. The ramp gain (74) associated with the two-state model ($K = 2$) was originally derived in [21] using linear regression and thereafter rederived in [17] in state space. The quadratic gain (75) for $K = 3$ and the cubic gain (76) for $K = 4$ have been originally derived and investigated in [17]. The gains (73)–(76) are reminiscent of the familiar Brown's exponential weight, although it can easily be verified that the latter is not unbiased.

Provided h_{li}, the relevant NG can also be found in an exact form. Using (41), (56), (63), and the property (59), one can find

$$\begin{aligned}
g_l &= \mathbf{W}_l^T \mathbf{W}_l = \sum_{i=0}^{N-1} h_{li}^2 \\
&= \sum_{i=0}^{N-1} h_{li} \sum_{j=0}^{l} a_{jl} i^j = \sum_{j=0}^{l} a_{jl} \sum_{i=0}^{N-1} h_{li} i^j \\
&= a_{0l}.
\end{aligned} \quad (77)$$

An identity $g_l = a_{0l}$ states that diminishing noise in the estimate means reducing the coefficient a_{0l} in the polynomial gain (63). By (77), the NGs associated with (73)–(76) can instantly be written as, respectively,

$$g_0 = \frac{1}{N}, \quad (78)$$

$$g_1 = \frac{2(2N-1)}{N(N+1)}, \quad (79)$$

$$g_2 = \frac{3(3N^2 - 3N + 2)}{N(N+1)(N+2)}, \quad (80)$$

$$g_3 = \frac{8(2N^3 - 3N^2 + 7N - 3)}{N(N+1)(N+2)(N+3)}. \quad (81)$$

Figure 11 sketches the functions (78)–(81). The shadowed area is the dead zone, which

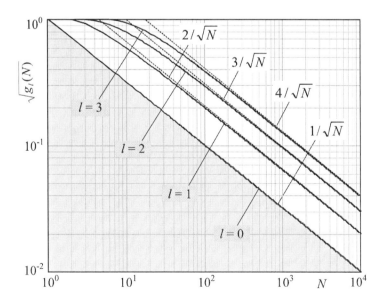

Figure 11. NG of the low-degree polynomial gains.

can never be reached. The bound of this zone is depicted by $g_0 = 1/N$ associated with simple averaging (73), because no one other filter is able to suppress noise so efficiently. For this reason, if $l > 0$, the function $\sqrt{g_l(N)}$ traces above the lower bound $1/\sqrt{N}$ and below the upper bound specified in [17] by

$$\frac{1}{\sqrt{N}} < \sqrt{g_l(N)} < \begin{cases} \frac{l+1}{\sqrt{N}}, & N \geqslant (l+1)^2 \\ 1, & N < (l+1)^2 \end{cases}, \quad l > 0. \tag{82}$$

It follows from (82) that the variance of the estimate noise basically behaves as a reciprocal of N and thus noise in the unbiased FIR estimate can be attenuated substantially when the allowed horizon N is large. Although it is a general rule, N cannot be infinite. That is owing to the aforementioned fact that the commonly used two-state and three-state clock models fit actual behaviors only on finite time intervals. The averaging horizon thus needs to be optimized.

3.3.3. Optimum Horizon and Sampling Interval

Returning back to Fig. 8, one can observe that, on the selected time interval on N points, the TIE behavior is well approximated with a quadratic function, the clock thus has 3 states, and the TIE can hence unbiasedly be filtered with the quadratic gain h_{2i}. On the other hand, one already knows that the noise variance in the estimate reduces as a reciprocal of N (Fig. 11). However, an attempt to increase N can make the estimate biased, because the order of the model may increase. The question thus arises of how to find the bound of the time interval over which the model acquires an additional state? At this bound, the MSE in the estimate obtained with h_{2i} would be minimum and the averaging horizon thus optimal, N_{opt}. Optimization must also be provided for the sampling interval τ, because N represents a number of τ on a horizon.

More specifically, Fig. 12 gives an idea about the estimate errors if we apply a ramp gain ($l = 1$) to a nonlinear TIE model ($K > 2$) on different horizons. For the illustrative purposes, we substitute the estimate with the goodness-of-fit linear test (dashed). On a short horizon $[n - N_1 + 1, n]$, the model is almost linear and the filter produces a negligible bias with, however, large noise, because N_1 is small (Fig. 11). With some N_2, both the bias and the noise are small. The value N_2 can thus be said to be optimal, N_{opt}, in the sense of the minimum MSE. With $N_3 > N_2$, the filter no longer fits the model with a large bias produced and despite the reduced noise.

To specify $N_{l\text{opt}}$ for the l-degree filter and the model with a redundant number of the states, $K > l + 1$, an analysis can be started as in [24] with the estimate (40), neglecting $\mathbf{N}_N$ for GPS-based measurements,

$$\hat{x}_n = \mathbf{W}_l^T \mathbf{C}_N \mathbf{x}_{n-N+1} + \mathbf{W}_l^T \mathbf{V}_N. \tag{83}$$

For (53), the MSE of (38) can be written as

$$\begin{aligned} J &= E\{[x_n - \mathbf{W}_l^T \mathbf{C}_N \mathbf{x}_{n-N+1} - \mathbf{W}_l^T \mathbf{V}_N]^2\} \\ &= (x_n - \mathbf{W}_l^T \mathbf{C}_N \mathbf{x}_{n-N+1})^2 + \mathbf{W}_l^T E\{\mathbf{V}_N \mathbf{V}_N^T\} \mathbf{W}_l \\ &= (x_n - \mathbf{W}_l^T \mathbf{C}_N \mathbf{x}_{n-N+1})^2 + \sigma_v^2 \mathbf{W}_l^T \mathbf{W}_l. \end{aligned} \tag{84}$$

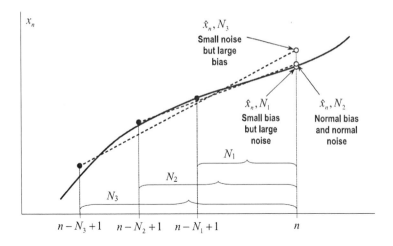

Figure 12. Errors of an unbiased ramp FIR filter at different horizons.

Table 1. Bias in the 1-state estimate.

l	bias		
	$K=1$	$K=2$	$K=3$
0	0	$-y\tau\frac{N-1}{2}$	$-\tau\frac{N-1}{2}\left(y + z\tau\frac{4N-5}{6}\right)$
1	0	0	$-z\tau^2\frac{(N-1)(N-2)}{12}$
2	0	0	0

It is seen that the bias in (84) is specified by

$$\text{bias}_n = x_n - \mathbf{W}_l^T \mathbf{C}_N \mathbf{x}_{n-N+1}. \tag{85}$$

If we further represent the model with K states and use the filter of the degree l, then the bias will depend on both K and l as shown in Table 1 [24]. Its value inherently reaches zero when $K \leqslant l+1$ and it is not zero if $K > l+1$. In turn, Table 2 gives the NG values calculated by $g_l = \sigma^2/\sigma_v^2 = \mathbf{W}_l^T \mathbf{W}_l$ with the approximations, by $N \gg 1$.

The MSE (84) may now be rewritten as $J = \text{bias}^2 + \sigma^2$ that, if we account for the dependencies on N and τ, becomes

$$J(\tau, N) = \text{bias}^2(\tau, N) + \sigma^2(\tau, N). \tag{86}$$

In order to find N_{opt} and τ_{opt}, the value $J(\tau, N)$ needs to be minimized.

General relations. To specify N_{opt} and τ_{opt} in the sense of the minimum MSE, one can first note that the state function $\mathbf{C}_N \mathbf{x}_{n-N+1}$ is defined on a horizon of N points, but

Table 2. Noise variance in the 1-state unbiased FIR estimate with $N \gg 1$.

l	Gain	$g_l = \sigma^2/\sigma_v^2$
0	Uniform	$\frac{1}{N}$
1	Ramp	$\left.\frac{2(2N-1)}{N(N+1)}\right|_{N\gg1} \cong \frac{4}{N}$
2	Quadratic	$\left.\frac{3(3N^2-3N+2)}{N(N+1)(N+2)}\right|_{N\gg1} \cong \frac{9}{N}$
3	Cubic	$\left.\frac{8(2N^3-3N^2+7N-3)}{N(N+1)(N+2)(N+3)}\right|_{N\gg1} \cong \frac{16}{N}$

is N-invariant for the given model. In contrast, the filter gain $\mathbf{W}_l$ is a function of N. To minimize (33) by N means thus solving for $N_{\mathrm{opt}}(\tau)$ an equation

$$\left.\frac{\partial}{\partial N}\mathrm{bias}^2(N)\right|_{N=N_{\mathrm{opt}}} + \left.\frac{\partial}{\partial N}\sigma^2(N)\right|_{N=N_{\mathrm{opt}}}$$

$$= 2[x_n - \mathbf{W}_l^T \mathbf{C}_N \mathbf{x}_{n-N+1}]\left.\frac{\partial \mathbf{W}_l^T}{\partial N}\right|_{N=N_{\mathrm{opt}}} \mathbf{C}_N \mathbf{x}_{n-N+1}$$

$$+ \sigma_v^2 \left.\frac{\partial g_l(N)}{\partial N}\right|_{N=N_{\mathrm{opt}}} = 0\,. \qquad (87)$$

On the other hand, the clock functions, x_n and $\mathbf{x}_{n-N+1}$, and the variance σ^2 are all τ-dependent. Similarly, to find τ_{opt}, one needs to solve

$$\left.\frac{\partial}{\partial \tau}\mathrm{bias}^2(\tau)\right|_{\tau=\tau_{\mathrm{opt}}} + \left.\frac{\partial}{\partial \tau}\sigma^2(\tau)\right|_{\tau=\tau_{\mathrm{opt}}}$$

$$= 2[x_n - \mathbf{W}_l^T \mathbf{C}_N \mathbf{x}_{n-N+1}]\left[\frac{\partial x_n}{\partial \tau} - \mathbf{W}_l^T \mathbf{C}_N \frac{\partial \mathbf{x}_{n-N+1}}{\partial \tau}\right]_{\tau=\tau_{\mathrm{opt}}}$$

$$+ g_l \left.\frac{\partial \sigma_v^2}{\partial \tau}\right|_{\tau=\tau_{\mathrm{opt}}} = 0 \qquad (88)$$

and then finally provide $N_{\mathrm{opt}}(\tau_{\mathrm{opt}})$.

In both cases, the partial derivatives can be taken either via the general relations or, for small l and K, using Tables 1 and 2. In the limiting case, albeit isolated, when the unbiased FIR filter fits the model over all N, the bias inherently become zero when $N_{\mathrm{opt}} \to \infty$ and τ_{opt} is derived from

$$g_l \left.\frac{\partial \sigma^2}{\partial \tau}\right|_{\tau=\tau_{\mathrm{opt}}} = 0\,. \qquad (89)$$

The latter would be illustrated by Fig. 12 if we suppose that the clock model is linear over all n. Because the bias in the clock models always depends on τ, we have only one more particular situation when the noise variance σ^2 is τ-invariant. This simplifies substantially a solution of (88). An additional look at Fig. 12 argues that, in order to find N_{opt} and τ_{opt}, it is enough to observe only such a part of the behavior, at which the MSE has a global minimum, thus choosing $N_3 < N < N_1$.

Optimal N_{opt} and τ_{opt} for low l and K. Below, we find $N_{\text{opt}}(\tau)$ for the low-state clock models, $1 \leqslant K \leqslant 3$, filtered with the low-degree FIR filters, $0 \leqslant l \leqslant 2$. It follows from Fig. 12 that $N_{\text{opt}}(\tau)$ is determined by increasing N for the given degree l of a filter (Table 1) and the $(l+2)$-state model. Because (87) is typically insoluble in simple functions for arbitrary N, we allow $N \gg 1$ suitable for GPS-based clock synchronization, thereby considering $N_{\text{opt}}(\tau)$ at a lower bound.

In the simplest case of simple averaging ($l = 0$) applied to a linear clock model ($K = 2$), the MSE (86) can be represented, using Table 1 and Table 2 for $N \gg 1$, as

$$
\begin{aligned}
J &= \text{bias}^2 + \sigma^2 \\
&= y^2\tau^2\frac{(N-1)^2}{4} + \left.\frac{\sigma_v^2}{N}\right|_{N\gg1} \\
&\cong y^2\tau^2\frac{N^2}{4} + \frac{\sigma_v^2}{N} .
\end{aligned}
\tag{90}
$$

Then testing (90) by (87) yields

$$
N_{\text{opt}}(\tau) = \sqrt[3]{\frac{2\sigma_v^2}{\tau^2 y^2}} ,
\tag{91}
$$

meaning that a horizon N is strongly restricted with the measurement noise variance σ_v^2, sampling interval τ, and the fractional frequency offset y of the clock oscillator. In fact, if measurement is noiseless, then $\sigma_v^2 = 0$ and the filter produces zero error even with a limiting horizon $N = 2$. If to increase the time step τ, then the number N inherently reduces. Finally, if the frequency offset y exists, then simple averaging is efficient only on small horizons with its 50% bias produced. On the other hand, if $y = 0$, then the TIE function has no slope and simple averaging can be applied at any horizon.

A similar result appears if we apply the ramp FIR filter ($l = 1$) to the three-state clock model ($K = 3$). As can be seen (Table 3 [24]), here the linear frequency drift rate z restricts a horizon in line with σ_v^2 and τ.

A substantially sophisticated situation arises when simple averaging is applied to the 3-state clock model or the ramp FIR filter to the 4-state model. Here, even by $N \gg 1$, (87) cannot be solved for $N_{\text{opt}}(\tau)$ in simple functions, although, several limiting cases can be recognized. For example, if we allow $y = 0$, then a solution for $l = 0$ and $K = 2$ can be found in Table 3.

If an instantaneous y is not available during the process, then the latter can be substituted with its maximum value featured to the particular clock. In this case, each of the functions listed in Table 3 represents the lower bound of $N_{\text{opt}}(\tau)$.

Remark: *The lower bounds of the optimal averaging horizons are provided by Table 3 if to substitute y and z with their maximum values featured to particular clocks.*

Table 3. Optimal horizons for the 1-state.

l	$N_{\mathrm{opt}}(\tau)$		
	$K = 1$	$K = 2$	$K = 3$
0	∞	$\sqrt[3]{\frac{2\sigma_v^2}{\tau^2 y^2}}$	$\sqrt[5]{\frac{9\sigma_v^2}{4\tau^4 z^2}}\ (y = 0)$
1	∞	∞	$\sqrt[5]{\frac{144\sigma_v^2}{\tau^4 z^2}}$
2	∞	∞	∞

The estimate variance via the GPS-based measurements can be supposed to be either τ-invariant or linearly growing with τ. Further testing (90) by (88) for these functions gives τ_{opt} zero or negative, respectively. Because neither of these values is practically available, one must take a minimum feasible one.

Remark: *The optimal sampling interval for GPS-based optimal FIR filtering of the clock TIE using 1 PPS signals is always $\tau_{\mathrm{opt}} = 1\ s$.*

Figure 13 shows the experimentally found values[2] of $N_{\mathrm{opt}}(\tau)$ and τ_{opt} for the ramp FIR filtering of the TIE of a local crystal clock embedded in the Stanford Frequency Counter SR620 via GPS-based measurement[3]. As can be seen in Fig. 13b, the optimal τ is exactly 1 s as stated theoretically. The analytically found function $N_{\mathrm{opt}}(\tau)$ (Table 3) also fits well the measurement (Fig. 13a) for y determined in postprocessing.

3.4. Unbiased FIR Filtering Algorithms

An important property of FIR filtering is that it can be applied to measurements of each of the states separately [7, 9]. Relying only on the TIE measurements, the unbiased FIR filtering algorithm can therefore be designed for the complete state space clock model as shown in [17], even with individual optimal horizons and sampling intervals [24].

3.4.1. Basic Algorithm

The FIR filtering algorithm applied to TIE measurements s_n of the three-state clock using commercially available multichannel GPS timing receivers can be designed [17] as shown in Fig. 14.

The 2-degree unbiased FIR filter with the gain (75) is applied to s_n in order to produce the estimate $\hat{x}_n$ of the TIE with the random error approximately calculated by the NG (80). Because the fractional frequency offset is the discrete-time derivative of the TIE,

---

[2] A special software for determining $N_{\mathrm{opt}}(\tau)$ and τ_{opt} was created by Jorge Muñoz-Diaz in his Master Thesis.

[3] A GPS-based measurement set was designed and measurements with and without the sawtooth have been provided by Luis Arceo-Miquel in his Master Thesis.

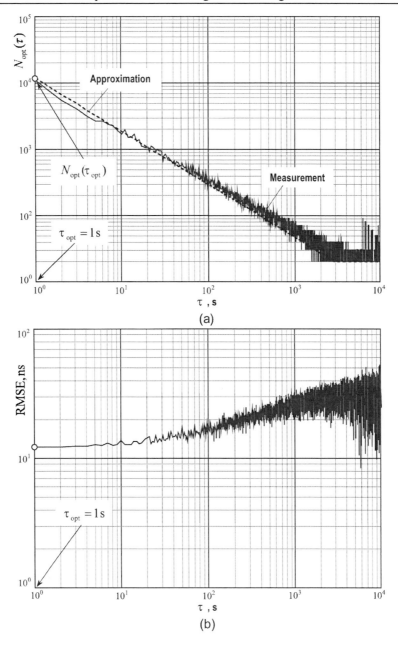

Figure 13. Measured optimal horizon and sampling interval for the ramp FIR filter: (a) $N_{\text{opt}}(\tau)$ and (b) τ_{opt} corresponding to the minimum root MSE (RMSE).

the measurement of y_n is further formed as $s_{yn} = (\hat{x}_n - \hat{x}_{n-1})/\tau$. Then the FIR filter with the ramp gain (74) is applied to s_{yn} to produce $\hat{y}_n$ with the variance calculated by the NG (79). The procedure is repeated using simple averaging (73) applied to the derivative $s_{zn} = (\hat{y}_n - \hat{y}_{n-1})/\tau$ of $\hat{y}_n$ to get an estimate $\hat{z}_n$ of the linear frequency drift rate z_n with the error calculated by the NG (78).

For the three-state and two-state clock models, the FIR filtering algorithm can thus be

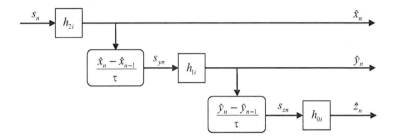

Figure 14. Basic unbiased FIR filtering algorithm for GPS-based measurements of the three-state clock model.

represented, in batch forms, as follows.

Three-state clock model. If the clock is identified to have 3 states, estimation of the states is provided with

$$\hat{x}_n = \sum_{i=0}^{N_x-1} h_{2i} s_{n-i}, \tag{92}$$

$$\hat{y}_n = \frac{1}{\tau} \sum_{i=0}^{N_y-1} h_{1i} [\hat{x}_{n-i} - \hat{x}_{n-j-1}], \tag{93}$$

$$\hat{z}_n = \frac{1}{\tau N_z} \sum_{i=0}^{N_z-1} [\hat{y}_{n-i} - \hat{y}_{n-i-1}], \tag{94}$$

where the averaging horizons, N_x, N_y, and N_z, are typically different for clock states. Inherently, the first accurate value of $\hat{x}_n$ appears at $N_x - 1$, of $\hat{y}_n$ at $N_x + N_y - 1$, and of $\hat{z}_n$ at $N_x + N_y + N_z - 1$.

Two-state clock model. For the two-state clock model, the existing states can be estimated with

$$\hat{x}_n = \sum_{i=0}^{N_x-1} h_{1i} s_{n-i}, \tag{95}$$

$$\hat{y}_n = \frac{1}{\tau N_y} \sum_{i=0}^{N_y-1} (\hat{x}_{n-i} - \hat{x}_{n-i-1}). \tag{96}$$

Here, the first accurate value of $\hat{x}_n$ appears at $N_x - 1$ and of $\hat{y}_n$ at $N_x + N_y - 1$.

3.4.2. Thinning Algorithm

Although the basic algorithm (Fig. 14) can efficiently be used for unbiased FIR filtering of clock models, it can be improved if we accept, as stated in [24], that optimal sampling intervals are not obligatorily 1 s for the second and third clock states. The improved algorithm called thinning was designed in [25] as shown in Fig. 15.

Allowed $\tau_x = 1$ s, the individual sampling intervals τ_y and τ_z are set for the second and third clock states, respectively. It is presumed that the relevant thinning coefficients, k_y and k_z, are found such that the principle properties of the clock states are not lost.

The first state x_n is estimated as in the basic algorithm by

$$\hat{x}_n = \sum_{i=0}^{N_x-1} h_{2i} s_{n-i} . \tag{97}$$

The function (97) is then thinned out by changing the sampling interval from τ to $\tau_y = k_y \tau$. In a new thinned time scale $m_y = \left\lfloor \frac{n}{k_y} \right\rfloor$, where the value $\left\lfloor \frac{a}{b} \right\rfloor$ represents an integer part of the ratio, the estimate (97) becomes $\hat{x}_{k_y m_y}$. Then the one pulse per k_y seconds (1PPk_yS) measurement of the second state is artificially formed by the backward time derivative as

$$s_{y k_y m_y} = \frac{\hat{x}_{k_y m_y} - \hat{x}_{k_y (m_y-1)}}{k_y \tau} . \tag{98}$$

The estimate $\hat{y}_{k_y m_y}$ of the second state is then found at each $k_y m_y$ point using the gain h_{1i} to yield

$$\hat{y}_{k_y m_y} = \sum_{i=0}^{N_y-1} h_{1i} s_{y k_y (m_y-1)} . \tag{99}$$

Similarly, the estimate $\hat{z}_{k_y k_z m_z}$ is found and we recognize two particular cases associated with the three-state and two-state clock models.

Three-state clock model. For the three-state clock model, the thinning unbiased FIR filtering algorithm gives us three estimates:

$$\hat{x}_n = \sum_{i=0}^{N_x-1} h_{2i} s_{n-i} , \tag{100}$$

$$\hat{y}_{k_y m_y} = \frac{1}{k_y \tau} \sum_{j=0}^{N_y-1} h_{1j} [\hat{x}_{k_y (m_y-j)} - \hat{x}_{k_y (m_y-j-1)}] , \tag{101}$$

$$\hat{z}_{k_y k_z m_z} = \frac{1}{k_y k_z \tau N_z} \sum_{r=0}^{N_z-1} [\hat{y}_{k_y k_z (m_z-r)} - \hat{y}_{k_y k_z (m_z-r-1)}] . \tag{102}$$

It can be shown that the first correct estimate of the third state, for the worst case of asynchronous thinning, appears at $N_x - 1 + k_y(N_y + k_z N_z)$.

Two-state clock model. The two-state model fits well atomic clocks and is often applied to crystal clocks with oven controlled crystal oscillators (OCXOs). The relevant thinning algorithm is shortened to

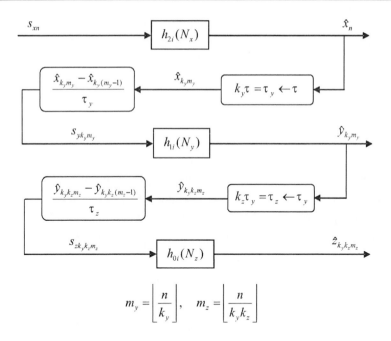

Figure 15. Thinning algorithm for unbiased FIR filtering of the clock states with individual horizons and sampling times.

$$\hat{x}_n = \sum_{i=0}^{N_x-1} h_{1i} s_{n-i}, \qquad (103)$$

$$\hat{y}_{k_y m_y} = \frac{1}{k_y \tau N_y} \sum_{j=0}^{N_y-1} [\hat{x}_{k_y(m_y-j)} - \hat{x}_{k_y(m_y-j-1)}]. \qquad (104)$$

In the worst case of asynchronous thinning, the first correct estimate of the second state appears at $N_x - 1 + k_y N_y$.

3.5. Applications to Crystal Clocks

In this section, the FIR filtering algorithms (Fig. 14 and Fig. 15) are applied to GPS-based measurements of the OCXO-based crystal clock embedded in the Stanford Frequency Counter SR620. Measurements are provided by the counter SR620 using the 1 PPS output, $\tau = 1$ s, of the GPS timing SynPaQ III Sensor. Simultaneously, to provide an actual trend and evaluate errors in the estimates, the TIE is measured by another counter SR620 using the Symmetricom cesium standard of frequency CsIII as a reference source. In some cases, the reference rubidium clock attached to the Stanford Frequency Counter SR625 is used. Initial time errors and frequency shifts in the reference sources are eliminated statistically in postprocessing.

At the early stage, the TIE model of the crystal clock was identified to have three-states, for which the algorithms (92)–(94) and (100)–(102) are applied. Optimum values of the horizons, N_x, N_y, and N_z, and the time steps, τ_y and τ_z, are found numerically, following [24], for the reference measurements in the sense of the minimum MSE.

In line with the FIR filter, the three-state Kalman filtering algorithm (Appendix A.) [15, 17, 25] is also aplied, to evaluate the trade-off. To use the Kalman filter, the uniformly distributed and delta-correlated measurement sawtooth noise is approximated with the Gaussian low having the variance $\Delta^2/3$ (11). The diffusion coefficients for the clock noise autocovariance matrix Ψ given by (14) are determined via the Allan variance as suggested in [13].

Noise in the states is evaluated in terms of the Allan deviation (ADEV) and time deviation (TDEV), respectively,

$$\sigma_y(\tau) = \sqrt{\frac{1}{2(M-2)\tau^2} \sum_{k=1}^{M-2} (x_{k+2} - 2x_{k+1} + x_k)^2} \qquad (105)$$

$$= \sqrt{\frac{1}{2(K-1)} \sum_{k=1}^{K-1} (y_{k+1} - y_k)^2}, \qquad (106)$$

$$\sigma_x(\tau) = \frac{\tau}{\sqrt{3}} \sigma_y(\tau) \qquad (107)$$

$$= \sqrt{\frac{1}{6(M-2)} \sum_{k=1}^{M-2} (x_{k+2} - 2x_{k+1} + x_k)^2}, \qquad (108)$$

where M is the number of time measurements and K is the number of frequency measurements.

3.5.1. The First Clock State

It was experimentally shown in [24] that, for the crystal clock imbedded in the SR620, the minimum MSE occurs at $\tau = 1$ s. For one day measurements with $\tau = 1$ s, the relevant optimal horizon is ascertained to be $N_x \cong 1550$ and the estimate $\hat{x}_n$ found using the quadratic gain (75), by (97).

Short base measurement. Figure 16a sketches a typical GPS-based measurement (for about 2 hours) as well as the FIR estimates $\hat{x}_n$ and the Kalman estimate $\hat{x}_{Kn}$. It is neatly seen that both estimates are consistent and unbiased. Also, they are noisy, as compared to the actual TIE behavior.

To figure out a picture of the noise more precisely, Fig. 16b sketches the TDEV (108) in both estimates obtained via one day measurements. Instantly, one indicates that noise in the unbiased FIR estimate is lower than in the Kalman estimate. In fact, the TDEV functions behave closely to each other, although a bit better performance in the range of up to $\tau = 2 \times 10^3$ exhibits the FIR filter.

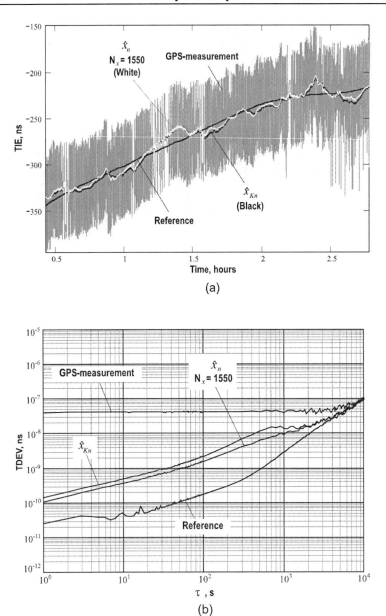

Figure 16. Typical short base GPS-based measurement and estimates of the TIE of a crystal clock having small frequency offset: (a) GPS-measurement and estimates and (b) TDEV.

Long base measurement. In the second experiment, the GPS-based measurements have been provided over one week and the optimal horizon was found to be $N_x \cong 5950$. Figure 17a illustrates the measurements along with the FIR and Kalman filtering estimates. In the region (I) in Fig. 17a, both filters fail owing to the GPS time errors. But, again, we watch here for a better performance of the FIR filter that is neatly reflected in the TDEV (Fig. 17b). Indeed, up to $\tau = 10^3$, the FIR filter produces lower noise, although, when

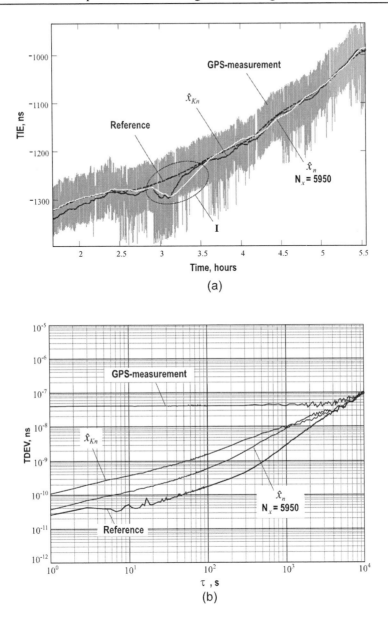

Figure 17. Typical one week GPS-based measurement and estimates of the TIE of a crystal clock: (a) measurement and estimates and (b) TDEV.

$10^3 < \tau < 5 \times 10^3$, the Kalman filter is a bit more preferable.

Figure 18 gives a typical picture for errors [16] that allows us to evaluate the trade-off between the unbiased ramp FIR filter and the two-state Kalman filter. First, we notice that there is no time delay between the estimates and the filters thus have equal time constants. It is also seen that the FIR filter is low sensitive to temporary uncertainties (demonstrates significantly smaller excursions), thus has better robustness than the Kalman filter. Yet, it

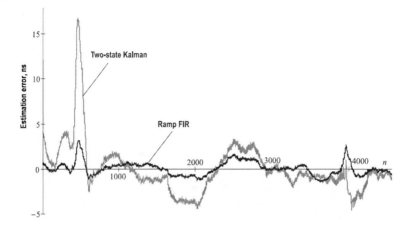

Figure 18. Typical estimation errors of the crystal clock 1-state via GPS-based measurements using the unbiased ramp FIR and two-state Kalman filters.

is seen that noise in the FIR filter is lower than in the Kalman one. Finally, let us notice that averaging featured to FIR filters inherently produces lower round-off errors than any recursive algorithm, including the Kalman filter.

3.5.2. The Second Clock State

For the second clock state, the minimum MSE in the unbiased FIR estimate was found in [25] to lie in the range of 10 s $\leqslant \tau \leqslant$ 100 s. Setting $k_y = 100$, an optimal horizon was found to be $N_y \cong 130$ and the estimation of y_n provided with the unbiased ramp FIR filter (74) and the Kalman filter (Appendix A.). Noise in both estimates was evaluated in terms of the ADEV (105).

Figure 19 exhibits typical results of the investigations. As can be seen (Fig. 19a), the FIR estimate $\hat{y}_n$ and the Kalman estimate $\hat{y}_{Kn}$ fit well both the reference measurement and the time derivative of the thinned out estimates of the TIE. As well as in the 1-state case (Fig. 17b), here the FIR filter produces much lower noise than the Kalman filter in the region up to about $\tau = 10^3$s. However, the Kalman filter has a bit better performance when 10^3 s $< \tau < 4 \times 10^3$ s.

3.5.3. The Third Clock State

The third clock state is known as the linear fractional frequency drift rate that is commonly associated with aging effects in the clock oscillator. Therefore, several days measurement is commonly not enough for accurate filtering of z_n. Instead, we can discuss the goodness-of-fit estimates associated with the FIR and Kalman filters. Fig. 20 gives an idea about accuracy of these filters.

As can be ascertained from Fig. 19a, the 2-state has both negative and positive time derivatives in the observed time interval. Hence, we expect that the 3-state should have both negative and positive estimates. An observation of Fig. 20 reveals that the FIR filter gives more or less correct values, whereas the Kalman filter produces only negative ones.

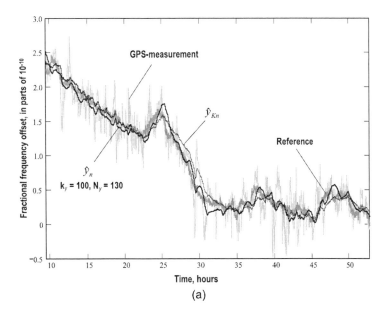

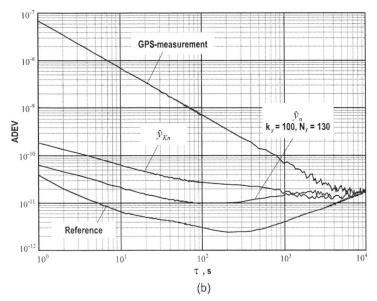

Figure 19. Measurements and estimates of the clock two-state: (a) FIR and Kalman estimates and (b) Allan deviation.

4. Prediction of Clock Instabilities

Prediction allows solving several critical problems in GPS-based timekeeping. It is necessary for the evaluation of possible future errors in the time scales formed by single and composed clocks. A 1-step prediction is used in the GPS-based receding horizon clock synchronization. The operation mode without temporary unavailable GPS synchronizing

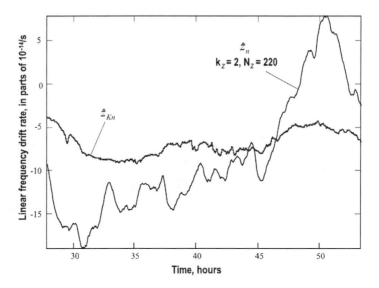

Figure 20. FIR estimate $\hat{z}_n$ and Kalman estimate $\hat{z}_{Kn}$ of the clock 3-state.

signals is known as "holdover." A key solution for holdover is the p-step ahead prediction that is often crucial for both ground and space applications.

In applications to clocks, prediction thus can be required either at the current point n from the past or at some $n + p$ point from the past and present. Below, we observe possible solutions allowed by FIR filtering.

4.0.4. State Space Representation on Finite Horizons

The predictive FIR filtering problem can be solved similarly to filtering, if we provide weighted averaging of the data taken from $n - N + 1 - p$ to $n - p$ with $p > 0$ and relate the result to the current point n as shown in Fig. 21a. The prediction at $n + p$ (Fig. 21b) can further be obtained by changing a variable.

Provided the recursive forward-in-time computation of $\mathbf{x}_n$ from $n - N + 1 - p$ to n, the state model (15) can be rewritten as

$$
\begin{aligned}
\mathbf{x}_n &= \mathbf{A}\mathbf{x}_{n-1} + \mathbf{w}_n \\
\mathbf{x}_{n-1} &= \mathbf{A}\mathbf{x}_{n-2} + \mathbf{w}_{n-1} \\
&\vdots \\
\mathbf{x}_{n-p} &= \mathbf{A}\mathbf{x}_{n-1-p} + \mathbf{w}_{n-p} \\
&\vdots \\
\mathbf{x}_{n-N+2-p} &= \mathbf{A}\mathbf{x}_{n-N+1-p} + \mathbf{w}_{n-N+2-p} \\
\mathbf{x}_{n-N+1-p} &= \mathbf{x}_{n-N+1-p} \,.
\end{aligned}
\tag{109}
$$

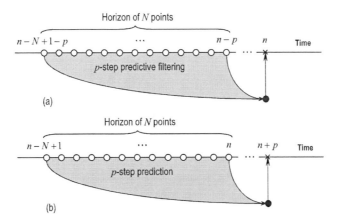

Figure 21. FIR prediction on a horizon of N points: (a) predictive filtering at n and (b) prediction at $n + p$.

Substituting the succeeding equations to the preceding ones of (109) gives

$$\begin{aligned}
\mathbf{x}_n &= \mathbf{A}(\mathbf{A}\mathbf{x}_{n-2} + \mathbf{w}_{n-1}) + \mathbf{w}_n \\
&= \mathbf{A}^2(\mathbf{A}\mathbf{x}_{n-3} + \mathbf{w}_{n-2}) + \mathbf{A}\mathbf{w}_{n-1} + \mathbf{w}_n \\
&= \mathbf{A}^{N-1+p}\mathbf{x}_{n-N+1-p} + \mathbf{A}^{N-2+p}\mathbf{w}_{n-N+2-p} \\
&\quad + \cdots + \mathbf{A}\mathbf{w}_{n-1} + \mathbf{w}_n, \\
\mathbf{x}_{n-1} &= \mathbf{A}(\mathbf{A}\mathbf{x}_{n-3} + \mathbf{w}_{n-2}) + \mathbf{w}_{n-1} \\
&= \mathbf{A}^2(\mathbf{A}\mathbf{x}_{n-4} + \mathbf{w}_{n-3}) + \mathbf{A}\mathbf{w}_{n-2} + \mathbf{w}_{n-1} \\
&= \mathbf{A}^{N-2+p}\mathbf{x}_{n-N+1-p} + \mathbf{A}^{N-3+p}\mathbf{w}_{n-N+2-p} \\
&\quad + \cdots + \mathbf{A}\mathbf{w}_{n-2} + \mathbf{w}_{n-1}, \\
&\vdots \\
\mathbf{x}_{n-p} &= \mathbf{A}^{N-1}\mathbf{x}_{n-N+1-p} + \mathbf{A}^{N-2}\mathbf{w}_{n-N+2-p} \\
&\quad + \cdots + \mathbf{A}\mathbf{w}_{n-1-p} + \mathbf{w}_{n-p}, \\
&\vdots \\
\mathbf{x}_{n-N+2-p} &= \mathbf{A}\mathbf{x}_{n-N+1-p} + \mathbf{w}_{n-N+2-p}, \\
\mathbf{x}_{n-N+1-p} &= \mathbf{x}_{n-N+1-p},
\end{aligned} \qquad (110)$$

By uniting a part of the equations (110) related to the horizon of N points from $n - N + 1 - p$ to $n - p$ in a short matrix form, we have

$$\mathbf{X}_N(p) = \mathbf{A}_N \mathbf{x}_{n-N+1-p} + \mathbf{B}_N \mathbf{N}_N(p), \qquad (111)$$

where the $3N \times 1$ clock state vector is given by

$$\mathbf{X}_N(p) = \begin{bmatrix} \mathbf{x}_{n-p}^T & \mathbf{x}_{n-1-p}^T & \cdots & \mathbf{x}_{n-N+1-p}^T \end{bmatrix}^T \qquad (112)$$

and the $3N \times 3$ matrix $\mathbf{A}_N$ is specified with (23) and (24).

The $3N \times 1$ clock noise vector is p-dependent,

$$\mathbf{N}_N(p) = \left[\mathbf{w}_{n-p}^T \, \mathbf{w}_{n-1-p}^T \cdots \mathbf{w}_{n-N+1-p}^T\right]^T \tag{113}$$

and the $3N \times 3N$ clock noise matrix $\mathbf{B}_N$ is given in two equal forms of (26) and (27). As well as in (20), the vector $\mathbf{x}_{n-N+1-p}$ is given in (109) and (110) exactly, although it is randomly-valued. Therefore, the noise vector $\mathbf{w}_{n-N+1-p}$ in (113) is always zero-valued.

Reasoning similarly to (29), one can also represent the measurement equation (16) in the past history as

$$
\begin{aligned}
s_n &= \mathbf{C}\mathbf{x}_n + e_n \\
s_{n-1} &= \mathbf{C}\mathbf{x}_{n-1} + e_{n-1} \\
&\vdots \\
s_{n-p} &= \mathbf{C}\mathbf{x}_{n-p} + e_{n-p} \\
&\vdots \\
s_{n-N+2-p} &= \mathbf{C}\mathbf{x}_{n-N+2-p} + e_{n-N+2-p} \\
s_{n-N+1-p} &= \mathbf{C}\mathbf{x}_{n-N+1-p} + e_{n-N+1-p}.
\end{aligned} \tag{114}
$$

Substituting the clock states taken from (110) to (114) gives us

$$
\begin{aligned}
s_n &= \mathbf{C}\mathbf{A}^{N-1+p}\mathbf{x}_{n-N+1-p} + \mathbf{C}\mathbf{w}_n + \mathbf{C}\mathbf{A}\mathbf{w}_{n-1} + \mathbf{C}\mathbf{A}^2\mathbf{w}_{n-2} \\
&\quad + \cdots + \mathbf{C}\mathbf{A}^{N-2+p}\mathbf{w}_{n-N+2-p} + e_n \\
s_{n-1} &= \mathbf{C}\mathbf{A}^{N-2+p}\mathbf{x}_{n-N+1-p} + \mathbf{C}\mathbf{w}_{n-1} + \mathbf{C}\mathbf{A}\mathbf{w}_{n-2} + \mathbf{C}\mathbf{A}^2\mathbf{w}_{n-3} \\
&\quad + \cdots + \mathbf{C}\mathbf{A}^{N-3+p}\mathbf{w}_{n-N+2-p} + e_{n-1} \\
&\vdots \\
s_{n-p} &= \mathbf{C}\mathbf{A}^{N-1}\mathbf{x}_{n-N+1-p} + \mathbf{C}\mathbf{w}_{n-1} + \mathbf{C}\mathbf{A}\mathbf{w}_{n-2} + \mathbf{C}\mathbf{A}^2\mathbf{w}_{n-3} \\
&\quad + \cdots + \mathbf{C}\mathbf{A}^{N-2}\mathbf{w}_{n-N+2-p} + e_{n-p} \\
&\vdots \\
s_{n-N+2-p} &= \mathbf{C}\mathbf{A}\mathbf{x}_{n-N+1-p} + \mathbf{C}\mathbf{w}_{n-N+2-p} + e_{n-N+2-p} \\
s_{n-N+1-p} &= \mathbf{C}\mathbf{x}_{n-N+1-p} + e_{n-N+1-p}.
\end{aligned} \tag{115}
$$

The measurement equation can now be written on a horizon of N points from $n - N + 1 - p$ to $n - p$ in a short matrix form as

$$\mathbf{S}_N(p) = \mathbf{C}_N\mathbf{x}_{n-N+1-p} + \mathbf{G}_N\mathbf{N}_N(p) + \mathbf{V}_N(p), \tag{116}$$

where the $N \times 1$ measurement vector of the TIE is

$$\mathbf{S}_N(p) = \left[s_{n-p} \, s_{n-1-p} \cdots s_{n-N+1-p}\right]^T, \tag{117}$$

the $N \times 3$ measurement matrix is specified by (32). The $N \times 1$ measurement noise vector has the form of

$$\mathbf{V}_N(p) = \left[e_{n-p} \, e_{n-1-p} \, \cdots \, e_{n-N+1-p} \right]^T, \tag{118}$$

and $\mathbf{G}_N$ is given by (34) and (35).

Like in the case of filtering, (36) and (37), the p-step dependent state space model can now be represented on a horizon of N past points from $n - N + 1 - p$ to $n - p$ with two equations

$$\mathbf{X}_N(p) = \mathbf{A}_N \mathbf{x}_{n-N+1-p} + \mathbf{B}_N \mathbf{N}_N(p), \tag{119}$$

$$\mathbf{S}_N(p) = \mathbf{C}_N \mathbf{x}_{n-N+1-p} + \mathbf{G}_N \mathbf{N}_N(p) + \mathbf{V}_N(p), \tag{120}$$

where the matrices $\mathbf{A}_N$, $\mathbf{B}_N$, $\mathbf{C}_N$, and $\mathbf{G}_N$ were earlier derived for filtering. One can observe that the sign in p does not affect (119) and (120). If $p = 0$, the equations become (36) and (37) and, by letting $p < 0$, one can use them for smoothing. The latter, however, is commonly not used in GPS-based timekeeping.

4.1. Optimal Predictive FIR Filtering of the TIE

Based upon (120) and utilizing N measurement points from $n - N + 1 - p$ to $n - p$, the predictive FIR filtering estimate at n is obtained as follows,

$$
\begin{aligned}
\tilde{x}_n &= \sum_{i=p}^{N-1+p} h_{li}(p) s_{n-i} \\
&= \mathbf{W}_l^T(p) \mathbf{S}_N(p) \\
&= \mathbf{W}_l^T(p) [\mathbf{C}_N \mathbf{x}_{n-N+1-p} + \mathbf{G}_N \mathbf{N}_N(p) + \mathbf{V}_N(p)],
\end{aligned} \tag{121}
$$

where $h_{li}(p)$ is the l-degree FIR filter gain dependent on p and the gain matrix is represented with

$$\mathbf{W}_l(p) = \left[h_{l0}(p) \, h_{l1}(p) \, \cdots \, h_{l(N-1)}(p) \right]^T. \tag{122}$$

To find the p-dependent gain matrix $\mathbf{W}_{l0}(p)$ in the sense of the minimum MSE in the predicted estimate, one can follow the procedure used for filtering. According to the approach, the MSE can be written as

$$
\begin{aligned}
J &= E(x_n - \tilde{x}_n)^2 \\
&= E\{x_n - \mathbf{W}_l^T(p) [\mathbf{C}_N \mathbf{x}_{n-N+1-p} + \mathbf{G}_N \mathbf{N}_N(p) + \mathbf{V}_N(p)]\}^2.
\end{aligned} \tag{123}
$$

The cost function (123) can then be minimized similarly to (42), using the orthogonality condition [18],

$$E\{x_n - \mathbf{W}_{l0}^T(p) [\mathbf{C}_N \mathbf{x}_{n-N+1-p} + \mathbf{G}_N \mathbf{N}_N(p) + \mathbf{V}_N(p)]\}$$

$$[\mathbf{C}_N \mathbf{x}_{n-N+1-p} + \mathbf{G}_N \mathbf{N}_N(p) + \mathbf{V}_N(p)]^T = 0. \tag{124}$$

Now, it needs projecting the values from the averaging horizon to the current point n in order to find x_n for (124). Extracting a deterministic part of the first row of the first equation in (110), we write

$$x_n = (\mathbf{A}^{N-1+p})_1 \mathbf{x}_{n-N+1-p}.$$ (125)

Substituting (125) to (124) gives

$$E\{(\mathbf{A}^{N-1+p})_1 \mathbf{x}_{n-N+1-p} - \mathbf{W}_{l0}^T(p)[\mathbf{C}_N \mathbf{x}_{n-N+1-p} + \mathbf{G}_N \mathbf{N}_N(p) + \mathbf{V}_N(p)]\}$$

$$\times [\mathbf{C}_N \mathbf{x}_{n-N+1-p} + \mathbf{G}_N \mathbf{N}_N(p) + \mathbf{V}_N(p)]^T = 0,$$ (126)

where, it is implied, the random processes are mutually independent and uncorrelated for all p; that is,

$$E\{\mathbf{x}_{n-N+1-p} \mathbf{N}_N^T(p)\} = E\{\mathbf{N}_N(p) \mathbf{x}_{n-N+1-p}^T\} = \mathbf{0},$$ (127)

$$E\{\mathbf{x}_{n-N+1-p} \mathbf{V}_N^T(p)\} = E\{\mathbf{V}_N(p) \mathbf{x}_{n-N+1-p}^T\} = \mathbf{0},$$ (128)

$$E\{\mathbf{V}_N(p) \mathbf{N}_N^T(p)\} = E\{\mathbf{N}_N(p) \mathbf{V}_N^T(p)\} = \mathbf{0}.$$ (129)

The noise covariance matrices are

$$\boldsymbol{\Psi}_N(p) = E\{\mathbf{N}_N(p) \mathbf{N}_N^T(p)\},$$ (130)
$$\boldsymbol{\Phi}_V(p) = E\{\mathbf{V}_N(p) \mathbf{V}_N^T(p)\}$$ (131)

and the mean square initial state at $n - N + 1 - p$ is

$$\mathbf{R}_0(p) = E\{\mathbf{x}_{n-N+1-p} \mathbf{x}_{n-N+1-p}^T\}.$$ (132)

By providing averaging in (126), one arrives at the exact optimal predictive FIR filter gain transpose matrix

$$\mathbf{W}_{l0}^T(p) = (\mathbf{A}^{N-1+p})_1 \mathbf{R}_0(p) \mathbf{C}_N^T$$
$$= \times [\mathbf{Z}_0(p) + \mathbf{Z}_N(p) + \boldsymbol{\Phi}_V(p)]^{-1}.$$ (133)

where $\mathbf{Z}_0(p) = \mathbf{C}_N \mathbf{R}_0(p) \mathbf{C}_N^T$ and $\mathbf{Z}_N(p) = \mathbf{G}_N \mathbf{R}_N(p) \mathbf{G}_N^T$. As can be seen, (133) instantly becomes (44), if to let $p = 0$.

Resume: *The optimal gain (133) is general for FIR filtering ($p = 0$), prediction ($p > 0$), and smoothing ($p < 0$) of the state space model (15) and (16).*

In (133) we meet a familiar problem with a computation of the inverse matrix $\mathbf{Z}^{-1}(p) = [\mathbf{Z}_0(p) + \mathbf{Z}_N(p) + \boldsymbol{\Phi}_V(p)]^{-1}$ that has $N \times N$ dimensions and must exist. As well as in the case of (46), the problem is circumvented if we allow large N.

4.1.1. Large Averaging Horizon $N \gg 1$

Referring to the analysis of (46) provided for $N \gg 1$, one can discard $\mathbf{Z}_N(p)$ and $\mathbf{\Phi}_V(p)$ in (133) and arrive at an important approximation

$$\mathbf{W}_{l0}^T(p) = (\mathbf{A}^{N-1+p})_1 \mathbf{R}_0(p) \mathbf{C}_N^T \left[\mathbf{C}_N \mathbf{R}_0(p) \mathbf{C}_N^T \right]^{-1} . \tag{134}$$

By multiplying the both sides of (134) with $\mathbf{C}_N \mathbf{R}_0(p) \mathbf{C}_N^T$,

$$\mathbf{W}_{l0}^T(p) \mathbf{C}_N \mathbf{R}_0(p) \mathbf{C}_N^T = (\mathbf{A}^{N-1+p})_1 \mathbf{R}_0(p) \mathbf{C}_N^T ,$$

and then discarding $\mathbf{R}_0(p) \mathbf{C}_N^T$, one arrives at the familiar unbiasedness constraint that is now p-dependent,

$$\mathbf{W}_{l0}^T(p) \mathbf{C}_N = (\mathbf{A}^{N-1+p})_1 , \tag{135}$$

in which the gain matrix $\mathbf{W}_{l0}(p)$ exists from $n - N + 1 - p$ to $n - p$. As well as (51), this constraint inherently does not depend on the initial conditions and noise. It then follows that the gain (134) can be used with any $\mathbf{R}_0(p)$, although such that the inverse in (134) exists.

> **Resume:** *The unbiasedness (or deadbeat) constraint (135) is general for unbiased FIR filtering ($p = 0$), prediction ($p > 0$), and smoothing ($p < 0$) with large $N \gg 1$.*

4.1.2. Predictive Estimate Variance

The payment for prediction is larger noise in the estimate, which variance can be found at n, similarly to (52), by analyzing the MSE

$$\begin{aligned} J(p) &= E(x_n - \tilde{x}_n)^2 \\ &= E\{(\mathbf{A}^{N-1+p})_1 \mathbf{x}_{n-N+1-p} \\ &\quad - \mathbf{W}_l^T(p)[\mathbf{C}_N \mathbf{x}_{n-N+1-p} + \mathbf{G}_N \mathbf{N}_N(p) + \mathbf{V}_N(p)]\}^2 . \end{aligned} \tag{136}$$

For the purely unbiased prediction, by embedding the unbiasedness (135) and accounting for the commutativity of

$$\mathbf{W}_l^T(p) \mathbf{G}_{Nk} \mathbf{N}_N(p) = [\mathbf{G}_{Nk} \mathbf{N}_N(p)]^T \mathbf{W}_l(p) ,$$

$$\mathbf{W}_l^T(p) \mathbf{V}_N(p) = \mathbf{V}_N^T(p) \mathbf{W}_l(p) ,$$

we thus have

$$
\begin{aligned}
\sigma_x^2(p) &= E\left[\mathbf{W}_l^T(p)\mathbf{G}_{Nk}\mathbf{N}_N(p) + \mathbf{W}_l^T(p)\mathbf{V}_N(p)\right]^2 \\
&= E\{\mathbf{W}_l^T(p)\mathbf{G}_{Nk}\mathbf{N}_N(p)\mathbf{W}_l^T(p)\mathbf{G}_{Nk}\mathbf{N}_N(p)\} \\
&\quad + E\{\mathbf{W}_l^T(p)\mathbf{V}_N(p)\mathbf{W}_l^T(p)\mathbf{V}_N(p)\} \\
&= \mathbf{W}_l^T(p)\mathbf{G}_{Nk}E\{\mathbf{N}_N(p)\mathbf{N}_N^T(p)\}\mathbf{G}_{Nk}^T\mathbf{W}_l(p) \\
&\quad + \mathbf{W}_l^T(p)E\{\mathbf{V}_N(p)\mathbf{V}_N^T(p)\}\mathbf{W}_l(p) \\
&= \mathbf{W}_l^T(p)\left[\mathbf{G}_{Nk}E\{\mathbf{N}_N(p)\mathbf{N}_N^T(p)\}\mathbf{G}_{Nk}^T\right. \\
&\quad \left. + E\{\mathbf{V}_N(p)\mathbf{V}_N^T(p)\}\right]\mathbf{W}_l(p). \\
&= \mathbf{W}_l^T(p)\left[\mathbf{G}_{Nk}\boldsymbol{\Psi}_N(p)\mathbf{G}_{Nk}^T + \boldsymbol{\Phi}_V(p)\right]\mathbf{W}_l(p). \quad (137) \\
&= \mathbf{W}_l^T(p)\left[\mathbf{Z}_N(p) + \boldsymbol{\Phi}_V(p)\right]\mathbf{W}_l(p). \quad (138)
\end{aligned}
$$

An important special case of (138) can be considered similarly to (54). If we let $e_n \cong v_n$ and recall that, in GPS-based clock synchronization, the measurement noise dominates, then the variance of the prediction can approximately be found as

$$
\begin{aligned}
\sigma_x^2(p) &= \mathbf{W}_l^T(p)\mathrm{diag}\underbrace{\left(\sigma_v^2 \; \sigma_v^2 \; \cdots \; \sigma_v^2\right)}_{N}\mathbf{W}_l(p) \\
&= \sigma_v^2\mathbf{W}_l^T(p)\mathbf{W}_l(p) \quad (139) \\
&= \sigma_v^2 g_l(p), \quad (140)
\end{aligned}
$$

where the p-step dependent NG is specified by

$$
g_l(p) = \mathbf{W}_l^T(p)\mathbf{W}_l(p) \quad (141)
$$

that can be used to characterize noise in the p-step ahead prediction with large measurement noise and $N \gg 1$.

4.2. Unbiased Predictive FIR Filtering

To derive the predictive unbiased FIR filter, one can proceed with the deadbeat constraint (135) that, similarly to (57), can be rewritten as

$$
\begin{aligned}
&\left[h_{lp}(p)\, h_{l(1+p)}(p) \; \cdots \; h_{l(N-1+p)}(p)\right]
\begin{bmatrix}
1 & \tau(N-1) & \frac{\tau^2(N-1)^2}{2} \\
1 & \tau(N-2) & \frac{\tau^2(N-2)^2}{2} \\
\vdots & \vdots & \vdots \\
1 & \tau & \frac{\tau^2}{2} \\
1 & 0 & 0
\end{bmatrix} \\
&= \begin{bmatrix} 1 & \tau(N-1+p) & \frac{\tau^2(N-1+p)^2}{2} \end{bmatrix}. \quad (142)
\end{aligned}
$$

By equating the components of the matrices in the both sides of (142) and accounting for the range of existence of the gain $h_{li}(p)$, from p to $N-1+p$, one arrives at the fundamental properties of the p-step dependent unbiased FIR filter gain [22, 26],

$$\sum_{i=p}^{N-1+p} h_{li}(p) = 1,\tag{143}$$

$$\sum_{i=p}^{N-1+p} h_{li}(p)i^u = 0, \quad 1 \leqslant u \leqslant l.\tag{144}$$

A short matrix form of (143) and (144) appears similarly to (60),

$$\mathbf{W}_{l0}^T(p)\mathbf{V}(p) = \mathbf{J}^T,\tag{145}$$

in which the p-step dependent $N \times (l+1)$ Vandermonde matrix is

$$\mathbf{V}(p) = \begin{bmatrix} 1 & p & p^2 \\ 1 & 1+p & (1+p)^2 \\ 1 & 2+p & (2+p)^2 \\ \vdots & \vdots & \vdots \\ 1 & N-1+p & (N-1+p)^2 \end{bmatrix}.\tag{146}$$

Following the derivation of (62), one can get a solution

$$\mathbf{W}_{l0}^T(p) = \mathbf{J}^T[\mathbf{V}^T(p)\mathbf{V}(p)]^{-1}\mathbf{V}^T(p)\tag{147}$$

that, it follows, serves for unbiased either prediction ($p > 0$) or filtering ($p = 0$), or even smoothing ($p < 0$) of the clock models.

Resume: *The gain (147) is general for GPS-based unbiased FIR filtering ($p = 0$), prediction ($p > 0$), and smoothing ($p < 0$) of the polynomial clock models..*

4.2.1. Polynomial Gain of the Predictive Unbiased FIR Filter

Reasoning along similar lines, as for filtering, we can also represent the gain of the predictive unbiased FIR filter with the l-degree polynomial as

$$h_{li}(p) = \sum_{j=0}^{l} a_{jl}(p)i^j.\tag{148}$$

For the degree $l = 2$, we thus have

$$h_{2i}(p) = a_{02}(p) + a_{12}(p)i + a_{22}(p)i^2.\tag{149}$$

Further substituting (149) to (147) and rearranging the terms lead to a linear matrix equation

$$\mathbf{J} = \mathbf{D}(p)\mathbf{\Upsilon}(p),\tag{150}$$

$$\begin{bmatrix} 1 \\ 0 \\ 0 \end{bmatrix} = \begin{bmatrix} d_0(p) & d_1(p) & d_2(p) \\ d_1(p) & d_2(p) & d_3(p) \\ d_2(p) & d_3(p) & d_4(p) \end{bmatrix} \begin{bmatrix} a_{02}(p) \\ a_{12}(p) \\ a_{22}(p) \end{bmatrix}\tag{151}$$

where, most generally, $\mathbf{J}$ is given by (67), yet

$$\mathbf{\Upsilon}(p) = \underbrace{\left[a_{0(K-k)}(p)\, a_{1(K-k)}(p)\, \cdots\, a_{(K-k)(K-k)}(p)\right]^{T}}_{K-k+1}, \tag{152}$$

and the p-dependent matrix $\mathbf{D}(p)$ is specified with

$$\mathbf{D}(p) = \mathbf{V}^{T}(p)\mathbf{V}(p) = \begin{bmatrix} d_0(p) & d_1(p) & \cdots & d_l(p) \\ d_1(p) & d_2(p) & \cdots & d_{l+1}(p) \\ \vdots & \vdots & \ddots & \vdots \\ d_l(p) & d_{l+1}(p) & \cdots & d_{2l}(p) \end{bmatrix}, \tag{153}$$

where the components are defined by the Bernoulli polynomials as [22, 26]

$$d_m(p) = \sum_{i=p}^{N-1+p} i^m, \quad m = 0, 1, \ldots 2l, \tag{154}$$

$$= \frac{1}{m+1} \left[B_{m+1}(N+p) - B_{m+1}(p)\right]. \tag{155}$$

The analytic solution of (150) with respect to the coefficients of the polynomial (148) is given by

$$a_{jl}(p) = (-1)^j \frac{M_{(j+1)1}(p)}{|\mathbf{D}(p)|}, \tag{156}$$

where $\mathbf{D}(p)$ is the determinant and $M_{(j+1)1}(p)$ is the minor of (153).

4.2.2. Properties of the p-step Dependent Gain

Summarizing, Table 4 lists the most critical properties of the p-step dependent gain of the unbiased FIR filter. These properties are inherent for any p and can easily be proved following the above considered case of filtering, $p = 0$. Note that, contrary to filtering, the power of the predictive gain, $p > 0$, is not equal to $h_{l0}(p)$. This is because the gain merely does not exist at zero.

4.2.3. A p-step Predictive Unbiased Ramp FIR Filter

Because of a lesser divergency with large p, linear predictors are often optimal or close to optimal in the prediction of clock instabilities [27]. For the linear prediction, $l = 1$, the ramp filter gain is p-step dependent,

$$h_{1i}(p) = a_{01}(p) + a_{11}(p)i, \tag{157}$$

and its coefficients can be found, by (156), to be

$$a_{01}(p) = \frac{2(2N-1)(N-1) + 12p(N-1+p)}{N(N^2-1)}, \tag{158}$$

GPS-based Optimal FIR Filtering and Steering of Clock Errors

Table 4. Properties of the p-dependent gain of the unbiased FIR filter.

	Property
Range of existence	$h_{li}(p) = \begin{cases} h_{li}(p), & p \leqslant i \leqslant N - 1 + p \\ 0, & \text{otherwise} \end{cases}$
Unit area	$\sum_{i=p}^{N-1+p} h_{li}(p) = 1$
Zero moments	$\sum_{i=p}^{N-1+p} h_{li}(p) i^u = 0, \quad 1 \leqslant u \leqslant l$
Power	$\sum_{i=p}^{N-1+p} h_{li}^2(p) = a_{0l}(p)$

$$a_{11}(p) = -\frac{6(N - 1 + 2p)}{N(N^2 - 1)}. \tag{159}$$

If $p = 0$, (157) degenerates to the familiar non predictive ramp gain (74). For $p = 1$, $p = 2$, and $p = 3$, the gain (157) becomes, respectively,

$$h_{1i}(1) = \frac{2(2N + 1) - 6i}{N(N - 1)}, \tag{160}$$

$$h_{1i}(2) = 2\frac{2N^2 + 9N + 13 - 3(N + 3)i}{N(N^2 - 1)}, \tag{161}$$

$$h_{1i}(3) = 2\frac{2N^2 + 15N + 37 - 3(N + 5)i}{N(N^2 - 1)}. \tag{162}$$

Note that the ramp gain (160) for $p = 1$ was originally derived and investigated in [23] using the Lagrange multipliers. The gains (161) and (162) were found in [22]. Figure 22 sketches (160)–(162) along with (74). An analysis reveals that p results in a higher slope of the gain $h_{li}(p)$ and an increase on $6p/N(N + 1)$ in both the positive and negative peak values.

The NG associated with (157) is given by

$$\begin{aligned} g_1(p) &= a_{10}(p) \\ &= \frac{2(2N - 1)(N - 1) + 12p(N - 1 + p)}{N(N^2 - 1)}. \end{aligned} \tag{163}$$

Figure 23 illustrates (163) for different steps p manifesting that prediction is achieved at increase of noise and at expense of stability. Indeed, if $2 \leqslant N \leqslant N_b$, where N_b is determined by solving $g_l(p) = 1$, the filter becomes inefficient, because the NG exceeds unity and the measurement noise is thus gained, when NG $\geqslant 1$. On the other hand, by $N \gg N_b$, the NG poorly depends on p and fits the asymptotic function

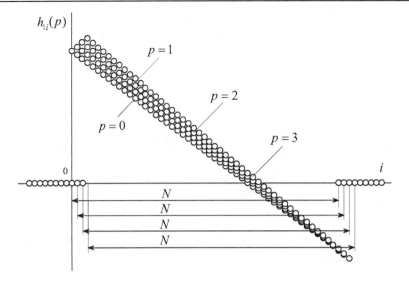

Figure 22. Ramp gains of the predictive unbiased FIR filter for small steps p. The case of $p = 0$ is not predictive.

$$g_1(p)|_{N \gg 1} \cong \frac{4}{N} \qquad (164)$$

that is dashed in Fig. 23. Moreover, when $N \ggg 1$, (161) makes the unbiased predictive FIR filtering estimate optimal in the sense of both zero bias and zero noise. This case, however, is rather theoretical, since no one real clock behaves linearly over all time and an infinite horizon thus cannot be applied.

4.2.4. A p-step Predictive Unbiased FIR Filter with a Quadratic Gain

Although the quadratic prediction is not commonly used, it can be applied when the clock behavior is more or less predictable on a short time base. Assuming $l = 2$ and an arbitrary step p, we write

$$h_{2i}(p) = a_{02}(p) + a_{12}(p)i + a_{22}(p)i^2 . \qquad (165)$$

The coefficients for (165) were found in [22] to be

$$a_{02}(p) = 3\frac{3N^4 - 12N^3 + 17N^2 - 12N + 4 + 12(N-1)(2N^2 - 5N + 2)p + 12(7N^2 - 15N + 7)p^2 + 120(N-1)p^3 + 60p^4}{N(N^2 - 1)(N^2 - 4)}, \qquad (166)$$

$$a_{12}(p) = -18\frac{2N^3 - 7N^2 + 7N - 2 + 2(7N^2 - 15N + 7)p + 30(N-1)p^2 + 20p^3}{N(N^2 - 1)(N^2 - 4)}, \qquad (167)$$

$$a_{22}(p) = 30\frac{N^2 - 3N + 2 + 6(N-1)p + 6p^2}{N(N^2 - 1)(N^2 - 4)}. \qquad (168)$$

For $p = 1$, $p = 2$, and $p = 3$, the function (165) degenerates to, respectively,

$$h_{2i}(1) = 3\frac{3N^2 + 3N + 2 - 6(2N + 1)i + 10i^2}{N(N-1)(N-2)}, \qquad (169)$$

$$h_{2i}(2) = \frac{3[3N^3 + 30N^2 + 125N + 146 - 2(6N^2 + 51N + 99)i + 10(N + 7)i^2]}{N(N^2 - 1)(N - 2)}, \qquad (170)$$

$$h_{2i}(3) = \frac{3[3N^4 + 60N^3 + 521N^2 + 1860N + 2308 - (12N^3 + 210N^2 + 1122N + 1860)i + (10N^2 + 150N + 380)i^2]}{N(N^2 - 1)(N^2 - 4)}, \qquad (171)$$

The quadratic gain (169) for $p = 1$ was originally derived in [23] via the Lagrange multipliers. The remaining gains, (170) and (171), can be found in [22].

Again, one notices that prediction is achieved at even larger increase of noise and at expense of stability. Figure 24 sketches $g_2(p) = a_{02}(p)$ for several small steps p. Observing

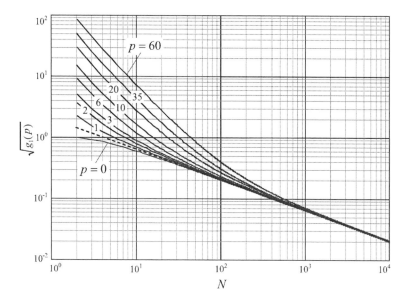

Figure 23. NG of the p-step predictive unbiased ramp FIR filter for several value of p. The case of $p = 0$ is not predictive. An asymptotic line (dashed) corresponds to $N \gg 1$.

Fig. 24, one can indicate that the quadratic prediction, first, becomes inefficient with much smaller p than in the ramp gain case (Fig. 23) and, second, a singularity occurs at $N = 2$ causing instability with small N. The range of singularity is shadowed in Fig. 24.

4.2.5. Generalizations

The following generalization can now be provided for a two-parameter family of the l-degree p-step predictive unbiased FIR filters:

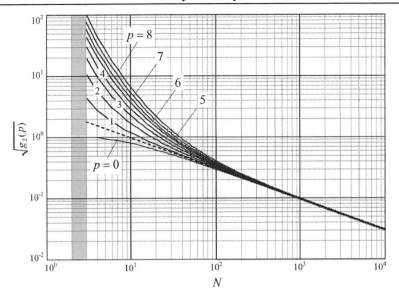

Figure 24. NG of the p-step predictive unbiased quadratic FIR filter. The case of $p = 0$ is not predictive. An asymptotic line (dashed) corresponds to $N \gg 1$. The range of singularity is shadowed.

- The range of instability in which $h_{li}(p)$ has singularities is $2 \leqslant N < l+1$. Note that N cannot be smaller than 2 in the averaging procedure. Beyond this range, the filter is addicted to instability as much as large is p.

- The filter is inefficient with $g_l(p) \geqslant 1$ (measurement noise is not attenuated) in the range of $l < N \leqslant N_b$, where N_b is determined by solving $g_l(p) = 1$.

- With large $N \gg N_b$, the NG fits an asymptotic function

$$g_l(N, p)|_{N \gg 1} \cong \frac{(l+1)^2}{N} \qquad (172)$$

dashed in Fig. 23 and Fig. 24. Note that the asymptotic function is p-invariant.

- If $p = N$, the NGs associated with the ramp and quadratic gain functions are calculated with high accuracy over all N by

$$g_1(p)|_{p=N} \cong \frac{28}{N}, \qquad (173)$$

$$g_2(p)|_{p=N} \cong \frac{873}{N}, \qquad (174)$$

meaning that the ramp filter becomes efficient when $p = N > 28$ and quadratic when $p = N > 873$.

- With extremely large $N \ggg 1$, the prediction noise variance becomes negligible, by (166), and the estimate thus optimal in the sense of both zero bias and zero noise.

4.3. Applications to GPS-based Measurements

As we already mentioned, prediction is used in GPS-based timekeeping to estimate future errors in local timescales and composed clocks, obtain steering of clock errors, and solve the holdover problem when the GPS timing signals are temporary not available. For all these needs, the algorithms can be designed using the general p-step predictive FIR filter gain.

4.3.1. Prediction Algorithms

To predict the future clock state over some time interval, the algorithms can be designed as shown in [26] and illustrated in Fig. 27 for holdover. Here the GPS-based measurement

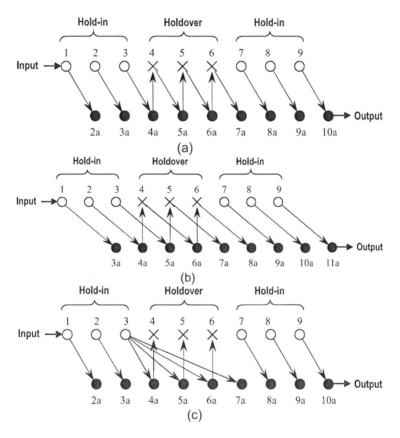

Figure 25. Prediction algorithms in applications to holdover: (a) fixed-step ($p = 1$), (b) fixed-step ($p = 2$), and (c) variable-step ($p = $ var).

(input) is depicted with "∘", the predicted estimate (output) with "•", and the unavailable or excluded "bad" measurement points by "×". It is implied that every measured point (from 1 to 9) represents the last point of the averaging horizon of N points shown in Fig. 21. From each of these points the predictive FIR filter (or the FIR predictor) is able to produce the p-step toward estimate. For example, the point 1 (Fig. 25a) representing a horizon of N neighboring past points, produces the estimate at 2a associated with the measurement at

2. The algorithm can be organized either with the fixed step (Fig. 25a and 25b) or variable step (Fig. 25c).

In the fixed-step case of $p = 1$ (Fig. 25a), the algorithm operates in "Hold-in" (from 1 to 3) producing the estimates (from 2a to 4a). At 4, the GPS-based measurement is supposed to be not available. Therefore, a predicted value (4a) is used instead and the filter produces the next estimate at 5a. Such a "Holdover" procedure is applied from 4 to 6. The filter then returns to "Hold-in". In a like manner, the algorithm is organized for the case shown in Fig. 25b and one can easily figure out how to design the fixed-step algorithm for any $p > 0$.

In the variable-step case illustrated in Fig. 25c, a horizon associated with the point 3 covers all the prediction range, producing individual estimates for the points 4a–7a. Here, the predicted values are not involved to the horizon and used in the following hold-in range.

4.3.2. Potential of FIR Prediction

Before processing real GPS-based measurement data, it is in order to discuss the predictive ramp FIR filter potential. To ascertain, a linear TIE function was simulated in [26] in the presence of discrete white Gaussian noise as shown in Fig. 26. Then the trade-off was shown between the fixed-step, $p = 1$ (Fig. 25a), and the variable-step, $p = $ var (Fig. 25c), algorithms. Both algorithms are run with $N = 500$ covering the holdover range of 2000 points. Ten predictive estimates for consequently generated random entries are shown in Fig. 26a and Fig. 26b for $p = 1$ and $p = $ var, respectively. Observing these results, one can infer that the prediction errors are similar.

Limiting errors of the fixed 1-step algorithm (Fig. 25a) were examined in [26] for the linear TIE function x_n caused by the fractional frequency offset of $y = 10^{-10}$. The TIE was generated in the presence of brightly pronounced uncertainty (rectangular jump) and the uniformly distributed random noise, representing the GPS receiver sawtooth, as shown in Fig. 27. To obtain holdover, with $p = 1$, the anomaly (rectangular jump) was excluded from the database. To provide filtering, with $p = 0$, the full database was processed.

Even a quick look at Fig. 27 assures us in a high efficiency of the solution. In fact, in the initial hold-in range, the values produced with $p = 0$ and $p = 1$ are indistinguishable. In the holdover range, the predicted estimate ($p = 1$) still does not get out of the actual linear behavior and keeps tracing it in the subsequent hold-in range. In turn, the filter ($p = 0$) exhibits inherent transients both within and beyond the holdover range.

4.3.3. Measurement with Sawtooth

In [26], the fixed-step prediction algorithm (Fig. 25a) has been examined using the ramp gain (157) and quadratic gain (165) for the GPS-based sawtooth measurement of the TIE. There has been used a crystal clock imbedded in the Stanford Frequency Counter SR620 (Stanford Research System, Inc., Sunnyvale, CA). The TIE was measured using the SR620 and the 1PPS signal of the SynPaQ III GPS Timing Sensor (Synergy Systems, LLC, San Diego, CA). The Symmetricom cesium standard of frequency CsIII was used as a reference time to measure actual x_n. To organize holdover, a part of the measurement is voluntary excluded from the database.

Figure 28 shows the predictive unbiased ramp FIR estimates for two steps, $p = 1$ and $p = 500$. The algorithm is applied with $N = 2000$. As can be seen, a real picture is similar

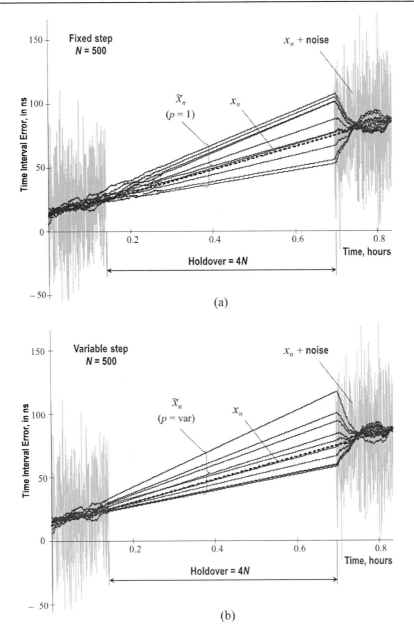

Figure 26. Typical unbiased FIR prediction (10 estimates for consequently generated random entries): (a) fixed-step, $p = 1$, and (b) variable-step, $p = $ var.

to that obtained by simulation (Fig. 27), although with larger errors caused by temporary measurement uncertainty. The other important point is that the prediction is highly insensitive to p when N is large. In fact, the difference between the predicted estimates produced by $p = 1$ and $p = 500$ results in Fig. 28 mostly in a time shift without substantial magnitude errors.

Fig. 29 demonstrates what happens if the predicted value is taken with a step p multiple

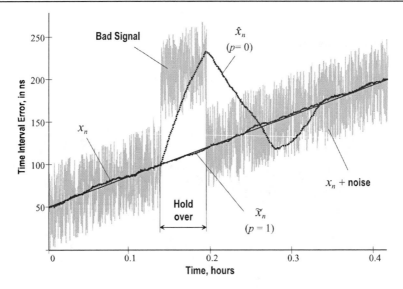

Figure 27. Typical 1-step, $p = 1$, unbiased ramp FIR prediction $\tilde{x}_n$ of a simulated linear TIE model in the presence of the uniformly distributed noise. The filtering estimate $\hat{x}_n$ is found for $p = 0$.

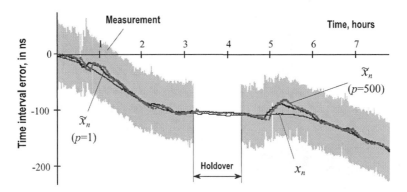

Figure 28. Unbiased predictive FIR estimates $\tilde{x}_n$ of the crystal clock TIE x_n, by $N = 2000$, $p = 1$, $p = 500$, and $l = 1$ via GPS-based measurements with sawtooth.

to 100 using the ramp predictor (Fig. 29a) and the quadratic predictor (Fig. 29b). As can be seen in Fig. 29b, the quadratic gain produces larger noise and the predicted estimate has greater divergency. This fact ensure us that linear predictors are optimal (or near optimal) in the GPS-based prediction of clock errors, as stated in [27].

4.3.4. Measurements with Sawtooth Correction

The experiment was repeated in [26] for the GPS-based measurements with sawtooth correction. A typical picture of the prediction is shown in Fig. 30. The result, however, does not reveal a substantially novelty. Indeed, the ramp FIR predictor still covers the holdover range with a sufficiently high accuracy and its estimate poorly depends on the step p. Moreover, the predicted values trace very closely to those provided in Fig. 28 via the sawtooth

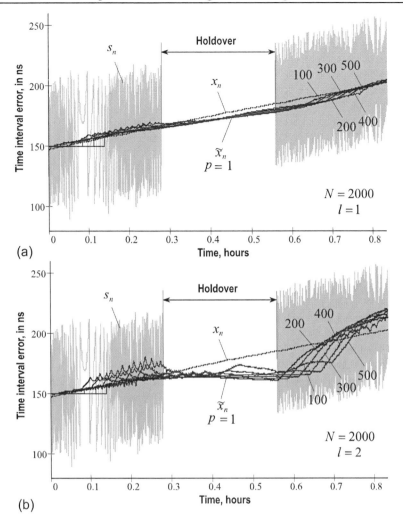

Figure 29. Unbiased FIR prediction $\tilde{x}_n$ of a crystal clock TIE x_n, for $N = 2000$ and different p: (a) ramp gain, $l = 1$, and (b) quadratic gain, $l = 2$.

measurements. This fact can be explained by the observation that the sawtooth correction works as an LP filter that eliminates sawtooth almost without touching the GPS time error. Because the optimal FIR predictor also suppresses the delta-correlated sawtooth noise and cannot eliminate efficiently the GPS error noise having a wide autocovariance function, the difference between the predicted estimates in Fig. 28 and Fig. 30 is poorly seen.

We notice that Figs. 27–30 sketch typical pictures for the unbiased FIR prediction via GPS-based measurements with and without the sawtooth. The worst prediction with a ramp gain can be figured out following Fig. 26.

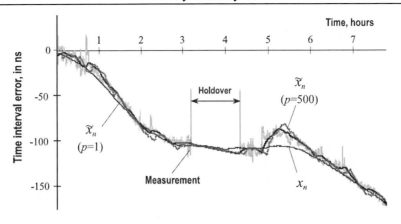

Figure 30. Unbiased predictive ramp estimates $\tilde{x}_n$ of the crystal clock TIE x_n, for $N = 2000$, $p = 1$, and $p = 500$ via the GPS-based sawtooth-less measurements.

5. Steering of Clock Errors

The need of steering local clocks and timescales arises with different allowed uncertainties in digital communications networks [28] as well as in bistatic radars, telephone networks [29], networked measurement and control systems [30], space systems, computer nets [31], etc. To enable it, the commercially available GPS timing receivers convey the reference time to the locked clock loop via the 1 PPS output. An organization of the loop is provided such that the local clock time error ranges over time below some allowed threshold that, for digital communication networks, is specified in [28] with a mask.

A generalized structure of GPS disciplining of a local clock is shown in Fig. 1. Reduction of clock errors can be achieved in two ways:

- Both the time error x_n and the fractional frequency offset y_n are adjusted via the GPS-disciplined oscillator (GPSDO).

- Only x_n is steered in the digital clock without touching a local oscillator.

In locked clocks, short term errors are owing to noise in a local oscillator that causes fundamental limits to a precision, whereas long term drifts are due to the GPS time that limits a clock accuracy. The effect is achieved using the Filter (Fig. 1), the design of which can vary. Averaging and low-order LP filters are typical for the commercially manufactured GPSDOs. It has been proposed in [32] to use averaging and integrating filters along with the phase locked loop (PLL). In [33], the use of the 3-order PLL is proposed in order to minimize phase errors. A linear least squares estimator is exploited in [34] to discipline a rubidium standard. Some authors propose applying the Kalman filters [15, 1] and even neural networks [35]. Experimental comparisons of several GPSDOs utilizing different filters can be found in [36, 37].

Below, we shall show that the problem can optimally be solved using the 1-step predictive unbiased FIR filter with the ramp gain (161). The p-step prediction algorithm (Fig. 25c) will be used for holdover, whenever the GPS timing signals are temporary unavailable.

5.1. GPS-based Synchronization Loop Model

The loop model of GPS-based synchronization of a local clock via the 1 PPS timing signals is shown in Fig. 1. The zero-mean GPS reference time error u_n is additively mixed with

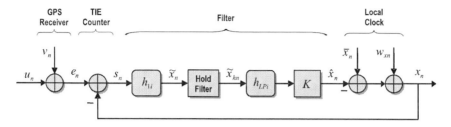

Figure 31. The loop model of GPS-based synchronization of a local clock using the 1 PPS timing signals.

the uniformly distributed sawtooth noise v_n induced by the receiver to form the zero-mean measurement noise e_n. In measurements with sawtooth correction, the component v_n is excluded from the model.

The clock average time error $\bar{x}_n$ is additively mixed with the zero-mean noise w_{xn}. To synchronize the clock with a highest accuracy, the white components of w_{xn} are commonly filtered out in the loop and the slowly changing flicker components are tracked along with $\bar{x}_n$ (Fig. 8). An averaging horizon is optimized in the sense of the minimum MSE.

The loop operates as follows. The clock TIE $x_n = \bar{x}_n + w_{xn}$ is subtracted from e_n by the time interval counter to produce the measurement $s_n = e_n - x_n$. It is implied that N such measurements are available in the nearest past history from $n - N$ to $n - 1$. To obtain a 1-step prediction $\tilde{x}_n$ at a current point n with a minimum divergency, the predictive unbiased FIR filter is used with the ramp gain $h_{1i}(p=1)$ given by (160). The predicted value $\tilde{x}_n$ then holds over time as $\tilde{x}_{hn}$ by the Hold Filter. The output of the hold filter is then smoothed with an LP filter having the impulse response h_{LPi} and gained with K to be $\hat{x}_n$. Finally, $\hat{x}_n$ is subtracted from x_n to yield the synchronization error $x_n = \bar{x}_n - \hat{x}_n + w_{xn}$. Provided normal functioning, the regular error $\bar{x}_n - \hat{x}_n$ is supposed to be near zero. The variance of the remaining random constituent of x_n is typically caused by the clock noise in a short time base and by the reference GPS timing signals in a long time base.

The main signals in the loop can be described as in the following. Suppose that s_n is measured from $n - N$ to $n - 1$. Then the predictive estimate $\tilde{x}_n$ of the clock TIE necessary for error steering in the next circle arrears at n as

$$\tilde{x}_n = -\sum_{i=1}^{N} h_{1i}(1) s_{n-i}, \qquad (175)$$

where the 1-step predictive ramp FIR filter (160) is represented with the gain

$$h_{1i}(1) = \begin{cases} \frac{2(2N+1)-6i}{N(N-1)}, & 1 \leqslant i \leqslant N \\ 0, & \text{otherwise} \end{cases}. \qquad (176)$$

As it has been discussed above, the gain (176) is unstable on short horizons, because $N = 1$ makes it infinite. This fact, however, does not mean too much for applications to clocks, because of large horizons used, $N \gg 1$. On the other hand, for holdover, (157) can be used with any step p.

To hold $\tilde{x}_n$ over time for steering x_n between the optimally defined values, the hold filter is used such that

$$\tilde{x}_{hn} = \tilde{x}_{\lfloor \frac{n}{M} \rfloor M} \,, \tag{177}$$

where $\lfloor \frac{n}{M} \rfloor$ is an integer part of n/M and M is a number of τ in the period of synchronization.

By multiple steering of clock errors with period τM, the hold filter produces a rectangular step signal. Applied straightforwardly to the clock, this signal obtains optimal step steering and assures that x_n ranges within the minimum bounds around zero.

For some applications, step steering may not be appropriate. Therefore, to smooth the output of the hold filter, an LP filter with the impulse response h_{LPi} is included. In what follows, we shall show experimentally that the 1-order LP filter with the impulse response

$$h_{LPi} = \begin{cases} Ae^{-\frac{\tau}{T}i} & , \quad i \geqslant 0 \\ 0 & , \quad i < 0 \end{cases}, \tag{178}$$

where T is the time constant and $A = 1 - e^{-\tau/T}$, is able to meed the demands of precise synchronization. Finally, if the LP filter is designed such that its input is not attenuated, one can set $K = 1$.

A supporting operation time diagram of the loop (Fig. 31) is given in Fig. 32. The FIR filter processes data from 0 to $N - 1$ producing a prediction $\tilde{x}_n$ at N (Fig. 32a). This value is held by the hold filter as $\tilde{x}_{hn}$ (Fig. 32c) and steers x_n (Fig. 32b). It is seen that x_n inherently jumps at $n = N$ closely to zero and thereafter, starting at $n = N$, behaves as in Fig. 32d (bold) following Fig. 32b. The function $\tilde{x}_{hn}$ is then smoothed (dashed), by an LP filter.

For multiple synchronization with period M, a smoothed value $\hat{x}_n$ is given, by (175)–(178) and the discrete convolution, as

$$\hat{x}_n = K \sum_{s=0}^{L-1} h_{LPs} \tilde{x}_{\lfloor \frac{n-s}{M} \rfloor M} \,, \tag{179}$$

where L is a reasonable length of the impulse response h_{LPi}.

Substituting (179) to $x_n = \bar{x}_n - \hat{x}_n + w_{xn}$ and accounting for (175) and $s_n = e_n - x_n$ allows us to write the loop general equation as

$$\begin{aligned} x_n = {} & \bar{x}_n - K \sum_{s=0}^{L-1} \sum_{i=1}^{N} h_{LPs} h_{1i}(1) x_{\lfloor \frac{n-s}{M} \rfloor M - i} \\ & + w_{xn} + K \sum_{s=0}^{L-1} \sum_{i=1}^{N} h_{LPs} h_{1i}(1) e_{\lfloor \frac{n-s}{M} \rfloor M - i} \,, \end{aligned} \tag{180}$$

where the difference between $\bar{x}_n$ and the first block of sums represents a regular synchronization error (bias) and the remaining term its random amount.

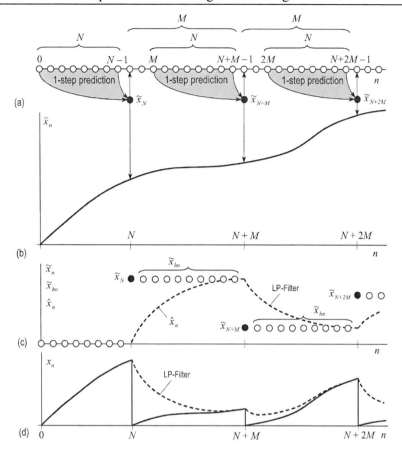

Figure 32. Time diagram of the GPS-based synchronization loop (Fig. 31).

5.2. Applications to GPS-locked Crystal Clocks

Below, as an example of applications, we examine a crystal clock with the oven controlled crystal oscillator (OCXO) embedded in the Stanford Frequency Counter SR620 and locked to the GPS timing signals using the SynPaQ III GPS Timing Sensor. Synchronization errors are evaluated in terms of the instantaneous error x_n, ADEV (105), and TDEV (108).

5.2.1. Instantaneous Synchronization Error

For GPS-based measurements with sawtooth, the optimum N has been experimentally found, following [24], to be $N_{\text{opt}} \cong 250$. By this value and without smoothing, $T = 0$, the TIE is obtained as shown in Fig. 33a. The effect produced by the LP filter with $T_{\text{opt}} \cong 2000$ is shown in Fig. 33b. One infers that, on the whole, the error in both cases ranges within the same bounds.

For sawtooth-less measurements, the optimal N has appeared to be $N_{\text{opt}} \cong 150$, leading to the results shown in Fig. 34a and Fig. 34b. As can be seen, the sawtooth correction does not change the picture cardinally, although it slightly draws together the TIE x_n bounds.

In order to investigate limiting errors, a special case was also considered below of the uncertainty-less steering with sawtooth and $N_{\text{opt}} \cong 250$. The effect is achieved by removal

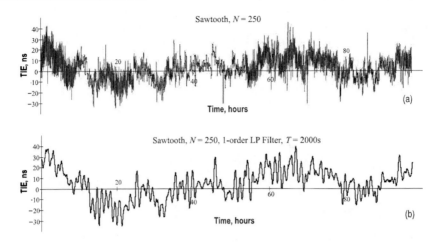

Figure 33. Typical time error x_n of the optimally GPS-locked crystal clock with sawtooth measurements: (a) step steering with $N = 250$ and (b) smoothed steering with the 1-order LP filter, $T = 2000$s.

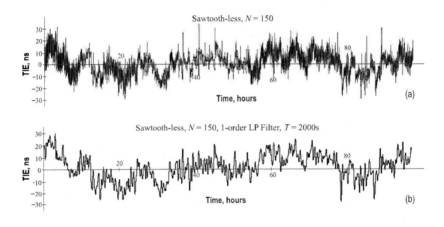

Figure 34. Typical time error x_n of the optimally GPS-locked crystal clock with sawtooth-less measurements: (a) step steering with $N = 150$ and (b) smoothed steering with the 1-order LP filter, $T = 2000$s.

of the GPS time error u_n from the measurement. The relevant synchronization errors are shown in Fig. 35 for steering without smoothing (a) and with the LP filter having $T = 2000$ s (b). Inherently, the TIE ranges very closely to zero at almost each of the points $n = N + mM$, unlike the case of both Fig. 33a and Fig. 34a. However, removal of the GPS time errors does not bring together substantially the error bounds. The synchronization error still fundamentally depends on the free time error of the unlocked clock (Fig. 1e), irrespective of noise in the reference signal. A comparison of Fig. 33b, Fig. 34b, and Fig. 35b assures us in this fact and we conclude that uncertainty in the GPS time is not the main limiter of the synchronization error.

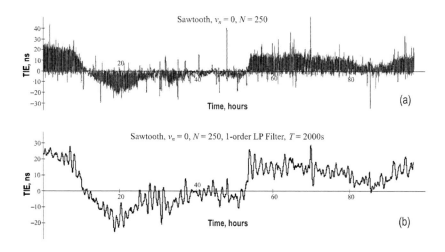

Figure 35. Typical limiting errors in the optimally GPS-locked crystal clock for measurements without uncertainties: (a) step steering with $N = 250$ and (b) smoothed steering with the 1-order LP filter, $T = 2000$s.

5.2.2. Allan Deviation

Noise in the locked crystal clock has been investigated in terms of the ADEV (105). The results are shown in Fig. 36. Here, "GPS 1 PPS, $v_n + u_n$" and "GPS 1 PPS, v_n" represent the ADEV of the 1 PPS signal with and without the sawtooth, respectively. "FIR Filter" represents the ADEV at the output of the optimal filter and "OCXO" the ADEV of the clock OCXO.

For the experimentally determined values of $N_{opt} = 250$ and $N_{opt} = 150$ associated with sawtooth and sawtooth-less measurements, respectively, Fig. 36 shows what happens with the ADEV if to change the time constant T of the LP filter.

Inherently, the ADEV of the 1PPS signal with sawtooth (Fig. 1c), traces upper than that without sawtooth (Fig. 1b). By the optimal FIR filter, the sawtooth noise reduces substantially (Fig. 36a), whereas the sawtooth-less measurement remains almost unaltered (Fig. 36b). In both cases, an LP filter forms the final picture. Namely, with small averaging times τ, the ADEV is due to noise in the unlocked clock and with large τ to noise in the output of the optimal FIR filter. The ADEV demonstrates an excursion, which peak value reaches a minimum with $T_{opt} \cong 2000$ s. The peak value displaces to the right, when $T > T_{opt}$, and to the left, if $T < T_{opt}$.

5.2.3. Time Deviation

Typical TDEV of the GPS locked crystal clock, calculated by (108), is illustrated in Fig. 37. For this measure, the optimal time constant of an LP filter must not only fit the minimum MSE, but also place the TDEV optimally below the mask specified in [28]. The relevant value has been found to be $T \cong 1000$ s, although, it can differ for other masks.

As can be seen, both the GPS-based sawtooth-less measurement and optimal FIR filter fit the mask [28] over almost all τ, whereas the sawtooth measurement almost does not fit.

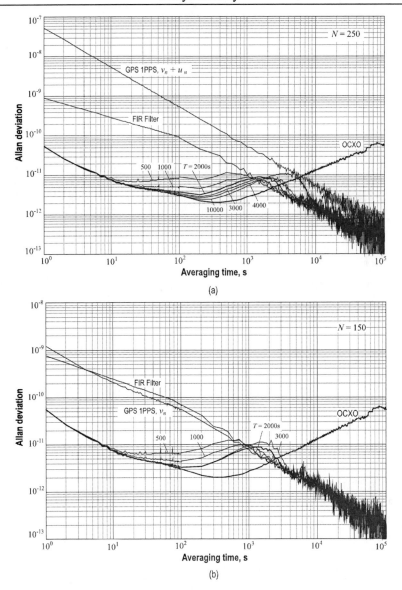

Figure 36. Typical ADEV of the GPS locked crystal clock for different T of the 1-order LP filter: (a) sawtooth measurement with $N_{opt} = 250$ and (b) sawtooth-less measurement with $N_{opt} = 150$.

A substantial improvement is achieved with the LP filter that places the TDEV much lower the mask. The latter means that the uselessness of the LP filter in the error bound reduction converts to the highly appreciated usefulness in reducing and optimizing the ADEV and TDEV with respect to some practically allowed mask.

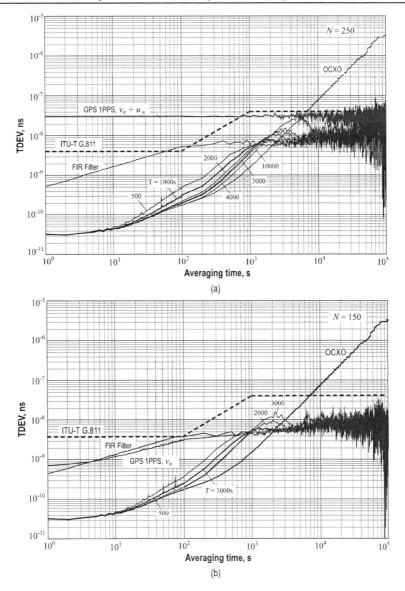

Figure 37. Typical TDEV of the GPS locked crystal clock for different T of the 1-order LP filter: (a) sawtooth measurement with $N_{\text{opt}} = 250$ and (b) sawtooth-less measurement with $N_{\text{opt}} = 150$.

5.2.4. Generalizations

Some important generalizations can now be provided regarding the GPS-based optimal steering of local clocks.

Effect of optimal FIR filter. It follows that the optimal FIR filter eliminates the sawtooth noise v_n similarly to the sawtooth correction and does not suppress substantially the GPS time error u_n. That means that the synchronization loop (Fig. 1) would operate with almost equal efficiency for the commercially available GPS timing receivers with and the

234 Yuriy Shmaliy

without sawtooth correction.

Optimum parameters. The minimum MSE in the locked crystal clock TIE can be obtained if to set the following optimized parameters:

- *Sampling time*: $\tau_{\mathrm{opt}} = 1$ s [24].

- *Averaging horizon*: It follows from Fig. 36a that N_{opt} lies at about the cross point of the ADEV power laws with the slopes $\tau^{-1/2}$ and τ^0.

- *Period of synchronization*: Although any period of synchronization M can be allowed, the filter horizon, by $M < N$, overlaps both the free and steered points, causing uncertainty errors. On the other hand, by $M > N$, synchronization is inefficient, because the points from $n = N$ to $n = M - N$ stay out of processing. The optimum value of M is thus $M_{\mathrm{opt}} = N_{\mathrm{opt}}$.

- *Time constant of an LP filter*: Depending on applications, the time constant T of an LP filter must be chosen to optimize the noise performance of a locked clock with respect to the practically required masks such as that specified in [28]. It follows that the optimum value T_{opt} lies at about the cross point of the ADEVs associated with the FIR Filter output and the unlocked crystal clock (OCXO).

Summarizing, Fig. 38 sketches a generalized picture for the error performance of an OCXO-based and GPS locked crystal clock. The ADEV of the GPS timing signals with sawtooth (A) has the slope τ^{-1} over all the observed range of τ. The same slope is featured to the GPS time error without sawtooth (B), although the function ranges lower. Optimal FIR filtering results in the sawtooth suppression such that the ADEV of the filter output traces closely to GPS-based measurements without sawtooth (B). The ADEV of a GPS locked crystal clock is due to the unlocked clock noise (C) with small τ and to the optimal filter output (B) with large τ.

To reach minimum possible errors (bold), it is recommended, first, finding N_{opt} at about the cross point between the ADEV power laws with $\tau^{-1/2}$ and τ^0 of the unlocked clock. By this value N_{opt}, the function (A) is lowered to (B). The goal would be reached if, second, to specify T_{opt} with the value found at about the cross point between (B) and (C). Let us notice that the ADEV excursion at the cross point between (B) and (C) cannot be avoided from the standpoint of control. It can only be minimized, by T_{opt}.

6. Conclusions

In this Chapter, we discussed GPS-based estimation and steering of the local clock time errors invoking methods of optimal FIR filtering. The approach has proved its usefulness owing to several important properties. In contrast to the recursive IIR structures, such as the Kalman filter, the transversal FIR ones are inherently BIBO stable and have better robustness against temporary uncertainties and round-off errors, especially for large N used in timekeeping. These properties has a special significance for the composed clocks, in which the prediction of future errors plays a key role and hence any error reduction is

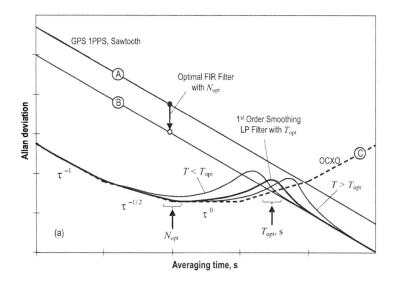

Figure 38. A generalization for the Allan deviation of the GPS locked crystal clock: (a) measurement with sawtooth, (b) sawtooth-less measurement, (c) unlocked clock with an OCXO, and (bold) locked clock.

highly appreciated. A disadvantage is an increased computational burden featured to large-dimension matrices associated with $N \gg 1$. The latter, however, is not crucial for the modern computer-aided processing, especially when FIR filtering is carried out in the iterative form [9, 10] and with the 1 s sampling time typically used in GPS-based timekeeping.

Designers would also appreciate that large averaging horizons substantially simplify the algorithm. Indeed, by large $N \gg 1$, the optimal FIR filter degenerates to a simple unbiased one relying on the polynomial gains having strong engineering features. That means, by extension, that there would be no reasonable necessity in using optimal filters to solve many of the clock problems. Simple unbiased FIR solutions can serve with a high efficiency for a variety of applications involving filtering, prediction, and smoothing. A number of practical examples given in this Chapter would certainly assure readers in this fact.

A. Kalman Algorithm

To evaluate the trade-off with the optimal FIR filter, the following three-state standard Kalman algorithm is applied to the state space model (15) and (16), whenever necessary. To run the algorithm, the clock noise components are approximated to be white Gaussian and the Chaffee's model (14) is used to describe the clock noise autocovariance matrix [13]. The measurement sawtooth noise is also approximated to be white Gaussian with the variance $V = \sigma_v^2$ given by (11).

For the initial estimate error $\mathbf{R}_{n-1}$ at $n-1$, the estimate error at n is predicted with

$$\tilde{\mathbf{R}}_n = \mathbf{A}\mathbf{R}_{n-1}\mathbf{A}^T + \mathbf{\Psi}$$

and the Kalman filter gain is calculated by

$$\mathbf{K}_n = \tilde{\mathbf{R}}_n \mathbf{C}^T (\mathbf{C}\tilde{\mathbf{R}}_n \mathbf{C}^T + V)^{-1}.$$

Thereafter, for the given state estimate $\hat{\mathbf{x}}_{n-1}$ at $n-1$, the Kalman estimate at n is provided by

$$\hat{\mathbf{x}}_n = \mathbf{A}\hat{\mathbf{x}}_{n-1} + \mathbf{K}_n(s_n - \mathbf{C}\mathbf{A}\hat{\mathbf{x}}_{n-1}).$$

The estimate error is then recalculated as

$$\mathbf{R}_n = (\mathbf{I} - \mathbf{K}_n\mathbf{C})\tilde{\mathbf{R}}_n$$

and the estimation process repeated recursively for the next time point.

References

[1] Shmaliy, Yu. S. *Continuous-Time Signals*; Springer: Dordrecht, 2006.

[2] Shmaliy, Yu. S. *Continuous-Time Systems*; Springer: Dordrecht, 2007.

[3] W. Lewandowski, W.; Petit, G.; C. Thomas. *IEEE Trans. Instrum. Meas.* 1993, 42, 2, 474–479.

[4] Meyer, F.; *Proc. 32nd Annu. PTTI Mtg.* 2000, 147–156.

[5] Allan, D. W.; Ashby, N.; Hodge, C. C. *The Science of Timekeeping*; Application Note 1289; Hewlett Packard: Englewood, CO, 1997, pp. 1–88.

[6] Kalman, R. E.; Bucy, R. S. *Trans. ASME J. Basic Engineering* 1961, 83D, 1, 95–108.

[7] Kwon, W. H.; Kim, P. S.; Park, P. *IEEE Trans. Automatic Control* 1999, 44, 9, 1787–1791.

[8] Fitzgerald, R. J.; *IEEE Trans. Autom. Control* 1971, AC-16, 736–747.

[9] Kwon, W. H.; Kim, P. S.; Han, S. H. *Automatica* 2002, 38, 3, 545–551.

[10] Kwon W. H.; Han, S. *Receding Horizon Control: Model Predictive Control for State Models*; Springer: London, 2005.

[11] Shmaliy, Yu. S.; Ibarra-Manzano, O.; Arceo-Miquel, L.; Muñoz-Diaz, J. *Sensors & Transducers J.* 2007, 79, 5, 1151–1156.

[12] *Standard definitions of physical quantities for fundamental frequency and time metrology - random instabilities*, IEEE Standard 1139; IEEE: Piscataway, NJ, 1988. pp. 1-20.

[13] Chaffee, J. W. *IEEE Trans. Ultrason., Ferroel., Freq. Contr.* 1987, 34, 6, 655–658.

[14] Allan, D. W.; Barnes, J. A. *Proc. 36th Annu. Freq. Contr. Symp.* 1982, 378–387.

[15] Stein, S. R.; Filler, R. L. *Proc. 42th Annu. Freq. Contr. Symp.* 1988, 447–452.

[16] Shmaliy, Yu. S. *IEEE Signal Proces. Letters,* 2008, 15, 517–520.

[17] Shmaliy, Yu. S. *IEEE Trans. Ultrason., Ferroel., Freq. Contr.* 2006, 53, 5, 862–870.

[18] Stark, H.; Woods, J. W. *Probability, Random Processes, and Estimation Theory for Engineers*; 2nd ed.; Prentice Hall: Upper Saddle River, NJ, 1994.

[19] Shmaliy, Yu. S.; Ibarra-Manzano, O. *Metrologia* (to be published).

[20] Kim, P.-S.; Lee, M.-E. *J. Computer Science and Technology* 2007, 22, 5, 779–784.

[21] Shmaliy, Yu. S. *IEEE Trans. on Ultrason., Ferroel. and Freq. Contr.* 2002, 49, 6, 789–797.

[22] Shmaliy, Yu. S. *Signal, Image, Video Process.*, doi 10.1007/s11760-008-0064-5. (in press)

[23] Heinonen, P.; Neuvo, Y. *IEEE Trans. Acoust. Speech Signal Process.* 1988, 36, 6, 892–899.

[24] Shmaliy, Yu. S.; Muñoz-Diaz, J.; Arceo-Miquel, L. *Digital Signal Process.* 2008, 18, 5, 739–750.

[25] Shmaliy, Yu. S.; Ibarra-Manzano, O.; Arceo-Miquel, L.; Muñoz-Diaz, J. *Measurement* 2008, 41, 5, 538–550.

[26] Shmaliy, Yu. S.; Arceo-Miquel, L. *IEEE Trans. Ultrason., Ferroel., Freq. Contr.* (to be published).

[27] Lepek, A. *Metrologia* 1997, 34, 379–386.

[28] *Timing characteristics of primary reference clocks*, ITU-T Recommendation G.811; ITU: Geneva, 1997.

[29] Pan, J. W. *IEEE Trans. on Ultrason., Ferroel. and Freq. Contr.* 1987, 34, 6, 629–638.

[30] *IEEE Standard 1588^T M-2002 for a precision clock synchronization protocol for networked measurement and control systems*; IEEE: Piscataway, 2002.

[31] Levine, J. *IEEE Trans. on Ultrason., Ferroel. and Freq. Contr.* 1999, 46, 4, 888–896.

[32] Helsby, N. C. *Proc. IEEE Int. Freq. Contr. Symp.* 2003, 435–439.

[33] Penrod, B. M. *Proc. IEEE Int. Freq. Contr. Symp.* 1996, 980–987.

[34] MacIntyre, A.; Stein, S. R. *Proc. 40th Annu. Freq. Contr. Symp.* 1986, 465–469.

[35] Chen, S. Y.; Yao, C. Y.; Xiao, G.; Ying, Y. S.; Wang, W. L. *Lecture Notes in Computer Science* 2005, 3512, 952–959.

[36] Davis, J. A.; Furlong, J. M. *Proc. 29th Annu. Precise Time and Time Interval (PTTI) Mtg.* 1997, 329–343.

[37] Lombardi, M. A.; Novak, A. N.; Zhang, V. S. *Proc. IEEE Int. Freq. Contr. Symp.* 2005, 677–684.

INDEX

A

academic, 83
access, 72, 98, 122
accounting, 71, 84, 90, 111, 112, 113, 114, 213, 214, 228
accuracy, viii, ix, x, 14, 29, 30, 34, 40, 44, 45, 53, 54, 55, 61, 62, 63, 64, 65, 66, 71, 72, 73, 74, 75, 77, 78, 83, 84, 91, 92, 93, 96, 100, 106, 111, 112, 113, 114, 124, 157, 164, 165, 171, 172, 175, 206, 220, 224, 226, 227
actuation, ix, 62, 85, 86, 90
actuators, 81, 82, 91
adaptation, 84
adaptive control, 64, 78, 79, 83, 85, 100, 101, 102, 103, 104, 105, 106, 109, 110, 111, 112, 113
adjustment, 49
adult, 58, 59
Aedes, viii, 57, 58, 59, 60
Africa, 118, 155, 156
aggregation, 77
aging, 186, 206
agricultural, viii, ix, 46, 61, 62, 63, 65, 67, 71, 77, 79, 81, 82, 84, 91, 93, 94, 96, 106, 112, 113, 114, 115, 116
agriculture, 91, 115
aid, vii, 131, 168
algorithm, ix, x, 8, 77, 78, 84, 86, 89, 98, 99, 102, 106, 109, 117, 129, 140, 171, 172, 198, 199, 200, 201, 202, 203, 206, 222, 226, 235
alluvial, 143
Alps, ix, 117, 118, 119, 120, 121, 140, 147, 152, 153, 154, 156
alternative, 15, 74, 112, 136, 172
alters, 78
ambiguity, 6, 68, 129, 168
amplitude, 94, 104
angular velocity, 69, 70
anomalous, x, 118, 142
ANOVA, 49

antenna, 2, 30, 31, 32, 33, 34, 35, 37, 46, 55, 65, 66, 72, 74, 75, 76, 78, 107, 111, 122, 124, 128, 130, 132, 133, 134
application, 27, 44, 55, 63, 82, 84, 85, 100, 108, 109, 111, 129, 130, 141
Arabia, 119
argument, 82
Asia, 147
Asian, 136, 146, 147
asphalt, 96
assignment, 129
assumptions, x, 4, 67, 70, 76, 140, 157, 158, 165, 168
asymptotic, 78, 82, 83, 98, 100, 111, 217, 219, 220
asymptotically, 83
asynchronous, 201, 202
atmospheric pressure, 130, 138
atomic clocks, 172, 175, 186, 187, 190, 201
autonomy, 31
availability, 21, 44, 129, 172
averaging, x, 171, 172, 179, 183, 184, 185, 186, 187, 188, 190, 192, 194, 197, 199, 200, 206, 208, 212, 220, 221, 226, 227, 231, 235

B

bandwidth, 73, 74, 86
batteries, 31
battery, 53, 122
behavior, x, 67, 77, 83, 86, 96, 97, 99, 100, 102, 104, 109, 110, 120, 136, 140, 150, 171, 172, 174, 179, 184, 189, 194, 197, 203, 218, 222
benchmark, 125, 126
benchmarks, 91, 94, 125
bending, vii, 1, 2, 3, 6, 7, 10, 13, 27
benefits, 28, 75
bias, 44, 54, 92, 129, 131, 138, 194, 195, 196, 197, 218, 220, 228
blocks, 118, 119, 120
boreal forest, 54, 55
boundary conditions, 160, 161, 163
boundary value problem, 160, 161, 163, 168, 169
bounds, 177, 178, 186, 197, 228, 229, 230

breeding, 59
Brownian motion, 139
buildings, 148

C

Canada, viii, 43, 46, 55
Carpathian, 119
carrier, vii, 1, 3, 4, 6, 7, 21, 33, 34, 35, 129
Cartesian coordinates, 128, 159
Caucasus, 155
cesium, 174, 202, 222
channels, 31, 33, 35, 172, 174
China, 171
civil engineering, viii, 29, 40
classical, 83
classification, 60
cleaning, 132
clock synchronization, 185, 197, 214, 237
closed-loop, 86, 114
closure, 44, 45, 46, 47, 53
codes, 26
collaboration, 40
Colorado, 28
commerce, vii
communication, 35, 129, 130, 132, 134, 135, 140, 141, 226
communities, 46
commutativity, 188, 213
compatibility, 159
compilation, 128, 154
complexity, 67, 118, 119
components, 14, 17, 84, 139, 140, 141, 142, 145, 158, 159, 161, 180, 181, 182, 183, 184, 185, 186, 187, 188, 189, 214, 216, 227, 235
computation, 15, 19, 73, 157, 186, 208, 212
computer systems, 30
computing, 6, 37, 102, 138, 164
concentration, 4, 168
confidence, 120, 145, 146
configuration, 68, 114, 119, 122, 128, 129, 157
coniferous, 44, 46
constraints, ix, 117, 129, 132, 133, 134, 135, 141, 142, 153, 155, 156
contamination, 32, 34
continuous data, 140
control, viii, ix, 57, 58, 59, 60, 61, 62, 63, 64, 65, 66, 67, 68, 71, 74, 77, 78, 79, 80, 81, 82, 83, 84, 85, 86, 87, 88, 89, 90, 91, 92, 93, 94, 95, 96, 97, 98, 99, 100, 101, 102, 103, 104, 105, 106, 108, 109, 110, 111, 112, 113, 114, 115, 116, 128, 132, 188, 226, 234, 237
convergence, 78, 81, 82, 83, 84, 91, 118, 119
conversion, 35
correlation, 32, 49, 52, 53, 175, 176
correlations, 140, 175
costs, 40
counterbalance, 113

coupling, 155
coverage, 9, 118, 142
covering, 222
CRC, 169
creep, 150
CRS, 122, 124
crust, ix, 118, 122, 136, 150, 158, 161, 168

D

data analysis, x, 122, 135, 157, 161, 168, 169
data collection, 30
data processing, 128, 152
data set, 135, 141, 142
database, 58, 59, 122, 124, 128, 222
decay, 9
decisions, 127
deficit, 150
definition, 30, 39, 70, 130, 133, 136, 155
deformation, ix, x, 117, 118, 119, 120, 121, 122, 124, 126, 127, 130, 132, 136, 139, 142, 146, 147, 149, 150, 153, 154, 155, 156, 157, 158, 161, 162, 163, 165, 168
demographic data, 59
dengue, viii, 57, 58, 59, 60
dengue hemorrhagic fever, viii, 57, 60
density, viii, 2, 8, 9, 10, 17, 18, 19, 27, 43, 47, 48, 118, 139, 149, 161, 174, 175, 177, 178
Department of Commerce, 28
Department of Defense, vii
deposits, 121
derivatives, 38, 164, 196, 206
designers, 63
detection, 134, 135
deviation, 2, 68, 77, 78, 81, 83, 84, 87, 90, 93, 94, 95, 97, 98, 99, 100, 101, 102, 103, 104, 105, 107, 108, 109, 110, 111, 174, 203, 207, 235
differential equations, 9
differentiation, 159, 162
diffusion, 179, 203
digital communication, 226
Dirac delta function, 175
direct measure, 72, 73, 74, 77
discipline, 226
discontinuity, 119
diseases, 59
displacement, x, 151, 157, 158, 161, 162, 164, 165, 168
distribution, viii, x, 2, 8, 38, 45, 49, 57, 58, 60, 71, 107, 121, 122, 139, 148, 157, 158, 161, 163, 164, 165, 168, 170
divergence, 136
division, 60, 119
Doppler, 28
duration, 90
dust, 62

Index 241

E

earth, 28, 59, 60, 130, 152
Earth Science, 117
earthquake, 132, 135, 142, 150, 151, 168
ecological, 44
ecology, 60
economic losses, 118
Education, 169
elasticity, 165, 169, 170
electric power, 122
electromagnetic, 28
electromagnetic wave, 28
electromagnetic waves, 28
electron, vii, 1, 2, 4, 6, 8, 9, 10, 11, 16, 17, 18, 19, 27
electron charge, 4
electron density, 2, 8, 9, 10, 17, 18, 19, 27
electron density distribution, 2, 8
electrons, 6, 9, 15, 16, 17
elk, 54
energy, 150
environment, 128, 186
environmental effects, 139
environmental factors, 185
epidemic, viii, 57, 58, 60
equating, 214
equilibrium, 79, 158, 159, 161, 165
equipment, ix, 61, 62, 63, 65, 66, 114
error estimation, 9, 135, 140
estimating, ix, x, 38, 49, 117, 135, 157
estimation process, 236
estimator, 174, 226
estimators, x, 171, 172, 188
Eurasia, 119, 128, 136, 162
Europe, 28, 118, 129, 131, 136, 145, 146, 147, 148, 149, 155, 171
evolution, 154, 178
examinations, 165
exclusion, 133

F

failure, 157
family, 172, 173, 177, 219
fatalities, 118
faults, 119, 121, 122, 149, 150, 152, 162
Federal Highway Administration, 41
feedback, 64, 82, 85, 90, 113, 114
Fermat, 3
fertilizers, 62
fever, 58, 60
fiber, 63, 115
filters, 74, 78, 189, 192, 197, 204, 205, 206, 219, 226, 235
finite element method, 164
FIR filters, 189, 192, 197, 206, 219
flooding, 60

flow, 86
FOG, 63
forests, 46
Fourier, 175
fragmentation, 119, 155
France, 91, 115, 116
freedom, 129, 131
friction, 150
funds, 54

G

Galileo, vii, 1, 2, 19, 27, 171
Gaussian, x, 171, 172, 174, 175, 177, 179, 180, 187, 188, 203, 222, 235
gene, 233
generalization, 219, 235
generalizations, 233
generation, 28, 137
generators, 122
Geneva, 237
Geographic Information System, viii, 29, 30, 37, 40, 41, 57, 58, 59, 60
geomagnetic field, 4, 8
geometrical parameters, 14
geophysical, 118
Germany, 1, 117, 136
Global Positioning System, vii, viii, ix, x, 1, 2, 3, 5, 7, 8, 9, 10, 11, 13, 14, 15, 17, 19, 21, 23, 25, 27, 28, 29, 30, 31, 32, 33, 34, 35, 36, 37, 39, 40, 41, 43, 44, 45, 46, 47, 49, 51, 53, 54, 55, 57, 58, 59, 60, 61, 62, 63, 65, 66, 72, 73, 74, 75, 76, 77, 78, 93, 97, 98, 104, 107, 111, 112, 113, 114, 115, 117, 118, 119, 120, 121, 122, 123, 124, 125, 126, 127, 128, 129, 130, 131, 132, 133, 134, 135, 136, 137, 138, 139, 140, 141, 142, 143, 147, 148, 149, 150, 151, 152, 153, 154, 155, 156, 157, 158, 159, 161, 162, 163, 164, 165, 167, 168, 169, 170, 171, 172, 173, 174, 175, 176, 177, 178, 180, 187, 188, 198, 202, 204, 207, 221, 222, 225, 226, 227, 229, 230, 231, 232, 233, 234, 235
goals, 135
goodness of fit, 132
graph, 93, 94
grassland, 46
gravity, 2, 106, 162
groups, 148
growth, 46
GSP, 34
guidance, viii, ix, 61, 62, 63, 64, 65, 66, 67, 71, 72, 73, 74, 75, 76, 77, 78, 79, 81, 82, 83, 84, 85, 86, 87, 88, 90, 91, 92, 93, 94, 95, 96, 97, 99, 100, 102, 103, 104, 105, 106, 109, 110, 111, 112, 113, 114, 115
guidelines, 154
gyroscope, 63

H

habitat, viii, 43, 44, 45, 47, 49, 52, 53, 54, 55
Hamiltonian, 9, 28
harvesting, 63, 82
hazards, 124
heat, 150
height, 9, 10, 14, 15, 17, 44, 47, 128, 133, 134, 137
hemisphere, 23, 47
Hermes, 115
high resolution, 173
high risk, 60
highways, 30, 35, 36, 38, 40, 41
histogram, 174, 175, 178
homeless, 118
horizon, 88, 89, 90, 114, 180, 182, 183, 184, 185, 186, 188, 194, 195, 199, 203, 204, 206, 207, 209, 210, 211, 212, 218, 221, 222, 227, 234
House, 57, 153
human, 157
human organisms, 157
humans, 58
hydrology, 138

I

identification, 60, 64, 86, 158
identity, 181, 182, 189, 190, 193
Illinois, 55, 63
imagery, 60
images, 9
implementation, 64, 141
incidence, 60, 94
incompressible, 163
independence, 53, 136
India, 171
indication, x, 118
indices, 159
industrial, 118
inelastic, x, 157
inertia, 63, 77, 78, 86, 90, 97, 106, 114
infection, viii, 57, 59
infectious, viii, 57, 60
infectious disease, viii, 57
infinite, 172, 177, 194, 218, 227
Information System, viii, 57, 58, 59
information systems, 41, 60
infrared, 63
infrastructure, 124
initial state, 182, 183, 185, 186, 212
innovation, 75
INS, 63
inspection, 131, 134
instabilities, 216, 236
instability, ix, 117, 219, 220
institutions, 148
instruments, 58

integration, 4, 5, 14, 16, 41, 160, 168
intensity, 118, 173
interaction, 118
interactions, 156
interface, 122, 131
interpretation, 155
interval, x, 34, 35, 38, 46, 49, 55, 126, 130, 171, 172, 173, 194, 197, 198, 199, 201, 206, 221, 227
inventories, 30
inversion, 81, 129, 130, 158, 159, 161, 162, 164, 165, 168, 170
ionization, 8, 9, 14, 15, 16, 17
ionosphere, 2, 3, 5, 6, 8, 9, 11, 14, 15, 17, 25, 26, 27, 28
ions, 62, 63, 65, 71, 72, 111
IRC, 69
isotropic, 159, 169
isotropy, 160, 168
Israel, 60
Italy, ix, 117, 118, 121, 152, 153, 154, 155, 156
iteration, 137

J

January, 153
Japan, x, 43, 45, 50, 51, 52, 53, 54, 55, 116, 157, 158, 161, 169, 170
Japanese, x, 55, 63, 155, 157, 158, 162, 163, 168, 170
Jurassic, 118

K

Kalman filter, 131, 133, 154, 172, 179, 203, 204, 205, 206, 226, 234, 236
Kalman Filtering, 154
Kenya, 60
kinematic model, ix, 61, 62, 64, 67, 68, 70, 71, 112, 114, 153
kinematics, 155
King, 129, 130, 132, 134, 135, 140, 141, 152, 154
knots, 38

L

Lagrange multipliers, 217
land, vii, 71, 96, 196
language, 66
larval, 60
law, ix, 8, 61, 62, 65, 67, 72, 73, 74, 78, 80, 81, 82, 83, 84, 85, 86, 87, 88, 89, 90, 91, 92, 93, 94, 95, 96, 97, 98, 99, 100, 101, 102, 103, 104, 105, 106, 108, 109, 110, 111, 112, 139, 174, 179
laws, ix, 61, 63, 64, 66, 67, 71, 75, 76, 77, 78, 79, 82, 91, 93, 96, 99, 100, 101, 103, 105, 106, 109, 110, 111, 112, 113, 114, 234

lead, 2, 25, 90, 102, 112, 114, 122, 128, 151, 185, 187, 191, 215
limitation, 85
limitations, x, 157, 161
linear, 7, 68, 69, 70, 72, 75, 79, 80, 81, 82, 86, 93, 112, 116, 129, 135, 159, 160, 161, 162, 163, 169, 174, 179, 180, 190, 191, 192, 194, 196, 197, 199, 206, 215, 216, 222, 224, 226
linear model, 79
linear regression, 192
linear systems, 116
link system, 122
lithosphere, 118, 119, 120, 153
localization, viii, 61, 62, 65
location, vii, viii, 23, 27, 30, 38, 41, 43, 44, 45, 46, 47, 48, 49, 50, 51, 52, 53, 54, 55, 68, 69, 72, 78, 92, 93, 112, 120, 125
logging, 35
London, 27, 154, 236
long distance, 148
longevity, 53
long-term, x, 118, 132, 150, 152
low power, 27
LTC, 173, 175

M

magnetic, viii, 2, 4, 8, 23, 28, 29, 31, 32, 40
magnetic field, 4, 8, 23, 28
magnetosphere, 27
magnets, 62
malaria, 60
management, 41
manipulation, 160
man-made, 60
manufacturer, 63, 65
mapping, 2, 30, 58, 130, 155
market, 63
marketing, 63
mask, 34, 35, 226, 231, 232
Massachusetts, 126, 154, 155
Massachusetts Institute of Technology, 115, 126, 129, 132, 134, 152, 154, 155
matrix, 127, 131, 179, 180, 181, 182, 183, 184, 185, 186, 188, 189, 190, 191, 203, 209, 210, 211, 212, 213, 215, 216, 235
MCS, 118
measurement, vii, 1, 2, 5, 8, 25, 26, 27, 34, 35, 38, 67, 72, 73, 74, 75, 76, 77, 78, 81, 88, 89, 93, 98, 99, 107, 108, 109, 110, 113, 115, 126, 139, 154, 164, 174, 175, 180, 182, 184, 185, 186, 187, 189, 198, 199, 201, 203, 204, 205, 206, 210, 211, 214, 217, 220, 221, 222, 223, 226, 227, 230, 231, 232, 233, 235, 237
measures, 2, 41, 158
median, 134
Mediterranean, ix, 117, 118, 119, 136, 152, 154, 155, 156

memory, 31, 34
meteorological, 59
Mexico, 54, 171
mining, ix
Ministry of Education, 54
mobile robot, 67, 79, 115
mobile robots, 67, 79, 115
model fitting, 184
modeling, 27, 40, 64, 67, 78, 96, 106, 113, 114, 128, 153
models, ix, x, 61, 62, 64, 67, 74, 79, 112, 113, 114, 130, 139, 155, 171, 191, 194, 196, 197, 199, 200, 201, 215
modulation, 179
modules, 127, 128
modulus, 168
mosquitoes, viii, 57, 58, 59, 60
motion, 63, 64, 69, 71, 72, 73, 78, 79, 91, 102, 112, 118, 121, 133, 134, 138, 139, 150, 152, 154, 155, 156, 161, 164
movement, 55, 59, 118, 136, 146, 148, 149, 150, 152, 162, 163

N

NAM, 162
national, 59
natural, 64, 80, 81, 82, 85, 112, 172
navigation system, 2
negative relation, 44, 52, 53
neglect, 187
Nepal, 154
network, 34, 35, 41, 122, 125, 127, 129, 132, 135, 140, 164, 169
neural network, 226
neural networks, 226
Nevada, 153
New Mexico, 54
New York, 28, 60, 169, 170
Nielsen, 54
NOAA, 60, 130
nodes, 164
noise, vii, ix, x, 1, 2, 25, 26, 27, 34, 98, 102, 117, 134, 135, 139, 140, 141, 142, 171, 172, 173, 174, 175, 176, 177, 179, 180, 181, 182, 183, 184, 185, 186, 187, 188, 189, 190, 192, 193, 194, 197, 203, 204, 206, 210, 212, 213, 214, 217, 218, 219, 220, 222, 224, 225, 226, 227, 230, 231, 233, 234, 235
non-linear, ix, 61, 62, 64, 74, 79, 80, 81, 82, 83, 84, 91, 92, 93, 95, 96, 97, 100, 101, 102, 103, 104, 105, 106, 109, 110, 111, 112, 113, 161, 174, 194
non-linearities, 82, 112
non-uniform, 168
normal, 99, 119, 139, 160, 174, 177, 227
North America, 54, 162
novelty, 224
numerical analysis, 158

O

observations, ix, 27, 117, 126, 131, 135, 138, 141, 155, 187
oil, 86
onchocerciasis, 60
online, 66, 72, 73, 77, 79, 84, 106, 112, 113
openness, 44, 47, 52
operating system, 33
optical, 2, 3
optimization, 90
orbit, vii, 2, 9, 10, 131, 139
organization, 226
orientation, ix, 54, 68, 117, 127, 134, 135, 141
orthogonality, 184, 211
oscillations, 73, 74, 77, 102, 107, 108, 110, 111, 112, 113
oscillator, 172, 173, 175, 197, 206, 226, 229
outliers, 134, 135, 137

P

Pacific, 162
paper, 2, 159
parameter, 38, 74, 79, 89, 103, 106, 133
Paris, 114, 115, 116
Parkinson, 28, 114, 115
partnership, 65
partnerships, 63
passenger, 30
perception, 63
performance, vii, viii, 17, 29, 30, 40, 43, 44, 45, 46, 47, 49, 50, 51, 52, 54, 55, 136, 203, 204, 206, 234
periodicity, 174
permittivity, 4
personal, 129, 130, 132, 134, 135, 140, 141
personal communication, 129, 130, 132, 134, 135, 140, 141
perturbation, ix, 62, 64, 67, 71, 72, 77, 78, 82, 83, 84, 87, 88, 98, 100, 109, 112, 113
perturbations, 74, 77, 82, 88, 102
pesticides, 62
PHS, 162
physics, 28
pitch, 72, 73, 74, 77, 113
plasma, 3, 4, 8
plastic, 161
plastic deformation, 161
Poisson, 160
polygons, 121
polynomial, x, 38, 171, 190, 191, 192, 193, 215, 216, 235
polynomial functions, 38
polynomials, 38, 216
poor, 53, 93, 102, 106
population, 58, 118
population density, 118

positive relation, 52, 53
positive relationship, 52, 53
power, 27, 122, 139, 175, 178, 179, 189, 216, 234
powers, 4, 5
PPS, 173, 175, 198, 202, 226, 227, 231
pragmatic, 26, 27
prediction, x, 75, 89, 90, 103, 114, 171, 186, 207, 208, 209, 213, 214, 215, 216, 217, 218, 219, 220, 221, 222, 223, 224, 225, 226, 227, 228, 229, 234, 235
predictors, 216, 224
pressure, 130, 138, 156
prevention, 59
preventive, viii, 57, 58
primary data, 131
probability, 54, 152, 177, 186
productivity, 62
program, 8, 10, 17, 122, 132, 135, 154
propagation, 3, 4, 8, 10, 13, 28, 113, 172
property, 192, 198
protocol, 237
pseudo, 129
public, viii, 57, 122, 126
public health, viii, 57
pulse, x, 171, 172, 201

R

radar, 60
radial distance, 15
radio, vii, viii, 1, 2, 3, 5, 8, 10, 27, 28, 43, 44, 45, 46, 47, 48, 49, 50, 51, 52, 53, 54, 155
radius, 47, 73
rainfall, 58
random, ix, x, 6, 117, 139, 140, 141, 171, 172, 173, 174, 175, 178, 179, 185, 198, 212, 222, 223, 227, 228
random errors, 6, 172
random walk, ix, 117, 139, 140, 141, 179
range, vii, 1, 2, 3, 4, 5, 6, 8, 10, 11, 13, 14, 21, 22, 23, 24, 27, 54, 103, 104, 106, 122, 129, 141, 148, 149, 150, 175, 203, 206, 214, 219, 220, 222, 224, 234
reading, 135
real time, 30, 122, 180
reality, 37
recall, 185, 188, 214
reception, 23, 62
reconstruction, viii, 29, 30, 40, 75, 99
recurrence, 152
reduction, 26, 27, 138, 232, 234
reference frame, ix, 68, 127, 133, 135, 136, 137, 141, 142, 143, 144, 145, 146, 147, 149, 154, 155
reference system, 135
refractive index, vii, 1, 3, 4, 5, 8
regional, 59, 124, 129, 131, 132, 133, 135, 154, 158, 161, 164, 165, 168
regression, 192

Index 245

regular, 129, 227, 228
rejection, 115
relationship, 44, 51, 52, 53, 54
relationships, viii, 43, 44, 45, 53
reliability, 44, 111
remote sensing, 60
rent, 88
repeatability, 138, 139
research, vii, 40, 44, 54, 58, 63, 65, 113, 169
research design, 54
researchers, viii, 30, 53, 61, 62, 122
residual error, 8, 11, 13, 15, 17, 20, 21, 26
residuals, 131, 132, 135, 141
resolution, 58, 129, 173, 178
returns, 98, 222
rheology, 119
Rift Valley fever, 60
rifting, 136
rigidity, 119
risk, 60
risk factors, 60
Robotics, 115
robustness, 172, 205, 234
rolling, 30, 64, 68, 70, 72, 74, 76, 83, 91, 99, 108, 111, 112, 113
Rome, 117
rotations, 69, 133
roughness, 38
rubidium, 186, 202, 226
rural, viii, 29, 30, 40
Russia, 171

S

safety, viii, 29, 40
sample, 73, 75, 77, 88, 89, 90, 124, 161
sampling, 65, 73, 75, 77, 86, 194, 197, 198, 199, 200, 201, 202, 235
SAR, 163
satellite, vii, viii, 1, 2, 3, 4, 6, 10, 14, 28, 34, 43, 44, 45, 49, 50, 51, 52, 53, 54, 60, 127, 129, 132, 133, 134, 139, 142
satellite orbits, 127
saturation, 82, 91
scalar, 75
scaling, 132, 135
Scandinavia, 136
scatter, 131, 132, 139
scattering, 130
scientists, 122
sea level, 36
searches, 37
security, 172
sediments, 140
seeds, 62
seismic, 118, 119, 120, 121, 124, 125, 149, 150, 152, 155
Senegal, 60

sensing, 60, 112, 113, 114
sensitivity, 186
sensors, 63, 114
separation, 14, 25, 87
series, 45, 122, 128, 130, 133, 134, 137, 138, 139, 140, 141, 142, 143, 144, 150, 152, 154, 155
services, 122
settlements, 118
SGA, 156
shape, ix, 45, 62, 89, 90
shear, 160, 162, 163, 164, 166, 167
shear deformation, 162, 163
shock, 107, 156
short period, 52
shoulders, 30
sign, 4, 16, 70, 72, 211
signals, vii, x, 2, 6, 7, 10, 11, 14, 15, 17, 28, 34, 75, 76, 138, 153, 171, 172, 180, 198, 208, 221, 226, 227, 229, 234
signal-to-noise ratio, 26
signs, 5
simulation, 24, 84, 85, 98, 99, 100, 109, 111, 112, 113, 223
simulations, 10, 23, 98, 99, 109
sine, 94, 95, 96
singular, 70
singularities, 220
sites, ix, 45, 46, 47, 49, 59, 117, 122, 123, 124, 125, 126, 127, 128, 130, 133, 134, 136, 137, 138, 139, 141, 142, 146, 147, 148, 152, 169
Slovenia, 156
smoothing, 38, 39, 211, 212, 213, 215, 229, 230, 235
software, 33, 35, 36, 37, 58, 59, 128, 130, 133, 140, 141, 198
solar, 2, 9, 11, 13, 14, 19, 122, 128
solar panels, 122
solutions, viii, ix, x, 29, 40, 117, 127, 130, 131, 132, 135, 139, 141, 142, 146, 148, 149, 156, 171, 180, 208, 235
South America, 154
Spain, viii, 29, 30, 36, 40, 115
spatial, 8, 41, 58, 59, 148, 161, 162, 164, 168
species, 46
spectral analysis, 73, 74, 140
spectrum, 132
speed, ix, 4, 5, 30, 35, 41, 86, 92, 118
speed of light, 4, 5
SPSS, 49, 55
stability, 82, 135, 189, 217, 219
stabilization, 115, 116, 133, 134, 135, 136, 137, 138, 141, 145
stabilize, 136, 139
stages, 118
standard deviation, 25, 26, 38, 73, 75, 92, 100, 103, 145, 146, 171, 172
standards, 171
statistics, 135
steady state, 75, 86
stochastic, 132

storage, 35
strain, x, 118, 122, 150, 151, 152, 153, 157, 158, 159, 160, 161, 162, 163, 164, 165, 166, 168, 170
stress, x, 118, 119, 150, 153, 157, 158, 159, 160, 161, 162, 163, 164, 165, 167, 168, 169, 170
summer, 46, 122
suppliers, 63
supply, ix, 61, 72, 122
suppression, 234
surveillance, vii, viii, 57, 58, 59, 60
synchronization, 172, 185, 186, 207, 227, 228, 229, 230, 233, 234
Synergy, 222
systems, vii, viii, x, 1, 2, 14, 30, 41, 58, 59, 60, 61, 62, 63, 79, 80, 112, 115, 171, 226, 237

T

technology, 41, 86, 122, 124
telephone, 226
temperature, 138, 174, 185
temporal, 140, 161, 168
Thailand, viii, 57, 58, 59, 60
theory, 27, 79, 190
third order, vii, 1, 2, 6, 7, 8, 10, 11, 14, 16, 19, 20, 21, 24, 27
three-dimensional, 9, 127, 135, 158
threshold, 226
tides, 130
TIE, x, 171, 172, 173, 174, 175, 178, 179, 180, 182, 184, 189, 191, 194, 197, 198, 202, 203, 204, 205, 206, 210, 211, 222, 224, 225, 226, 227, 229, 230, 234
time, vii, viii, ix, x, 6, 30, 41, 43, 44, 45, 49, 50, 51, 52, 53, 62, 76, 77, 79, 80, 81, 82, 86, 88, 90, 98, 99, 102, 111, 112, 115, 117, 122, 128, 130, 132, 133, 134, 137, 138, 139, 140, 141, 142, 143, 144, 150, 152, 153, 154, 155, 162, 171, 172, 173, 174, 175, 176, 177, 178, 179, 180, 185, 186, 187, 188, 189, 194, 201, 202, 203, 204, 205, 206, 207, 211, 218, 221, 222, 223, 225, 226, 227, 228, 230, 231, 232, 233, 234, 235, 236
time series, 122, 128, 133, 134, 137, 138, 139, 140, 141, 142, 143, 144, 150, 152, 154, 155
time use, 122
timing, x, 171, 172, 173, 175, 176, 177, 180, 187, 198, 202, 221, 226, 227, 229, 233, 234
Tokyo, 43, 157
tolerance, 38, 137
torque, 71
tracking, vii, 41, 44, 127, 189
traction, 159, 160, 162, 163
trade-off, 203, 205, 222, 235
traffic, viii, 29, 30, 35, 40, 41
trajectory, 81, 89, 90, 92, 93, 94, 95, 96, 100, 112
transformation, 81
transformations, 81, 185
transition, 38, 87, 150

translation, 69, 135, 137
transmission, 58, 59, 60, 85
transparent, 160
transportation, 41
transpose, 185, 190, 212
travel, 2, 6, 11, 30
travel time, 30
trees, 30
trend, 179, 202
trial, 168
triggers, 161
Turkey, 155
two-dimensional, 8, 158, 159
two-state model, 190, 192, 201

U

UHF, 66
uncertainty, x, 27, 131, 138, 139, 171, 172, 173, 174, 222, 223, 230, 234
uniform, x, 157, 159, 163, 164, 168, 175, 190, 192
United States, vii, 172
universities, 63
urban areas, 58

V

vacuum, 4
Valencia, 115
validation, 97, 106
validity, 168
valley fever, 60
values, 5, 9, 10, 13, 14, 17, 20, 21, 23, 38, 47, 49, 64, 72, 73, 74, 77, 83, 85, 87, 88, 89, 90, 93, 94, 95, 97, 98, 100, 102, 104, 107, 108, 112, 113, 114, 129, 132, 133, 134, 135, 137, 141, 151, 164, 177, 195, 197, 198, 203, 206, 212, 217, 222, 224, 228, 231
variable, 2, 44, 47, 49, 68, 71, 74, 79, 80, 83, 84, 85, 88, 90, 208, 222
variables, 67, 72, 74, 77, 88, 94, 113, 114
variance, 49, 175, 176, 177, 180, 185, 186, 187, 188, 189, 194, 196, 197, 198, 199, 203, 213, 214, 220, 227, 235
variation, 45, 104, 155
vector, viii, 4, 38, 57, 58, 59, 60, 68, 69, 72, 73, 77, 78, 83, 84, 87, 88, 112, 159, 160, 180, 181, 182, 183, 186, 209, 210
vegetation, 44, 45, 46, 47, 52, 60
vehicles, viii, ix, 29, 30, 35, 61, 62, 63, 64, 65, 67, 71, 75, 77, 78, 83, 85, 87, 91, 94, 96, 103, 104, 105, 106, 107, 109, 112, 113, 114, 115, 116
velocity, vii, ix, x, 4, 5, 6, 65, 68, 69, 70, 71, 72, 73, 74, 75, 79, 80, 81, 82, 83, 91, 92, 93, 94, 97, 100, 102, 105, 106, 109, 112, 117, 118, 122, 128, 135, 136, 138, 139, 141, 142, 146, 147, 148, 149, 150, 151, 152, 153, 155, 156, 168, 170

venue, 122
vessels, 71
village, 58, 59, 60
virus, viii, 57, 58, 60
visible, 9
vulnerability, 118

W

water, 30
wave propagation, 8
wavelengths, 155
web, 122, 126
WHO, 60
wildlife, 44

wind, 122
windows, 55
winter, 46
wireless, x, 171
wires, 86

Y

yield, 122, 170, 201, 227

Z

Zen, 130